Techniques for
Image Processing and Classification
in Remote Sensing

Academic Press Rapid Manuscript Reproduction

Techniques for
Image Processing and Classification
in Remote Sensing

ROBERT A. SCHOWENGERDT

Office of Arid Lands Studies
University of Arizona
Tucson, Arizona

ACADEMIC PRESS, INC.

(Harcourt Brace Jovanovich, Publishers)

Orlando   San Diego   New York   London

Toronto   Montreal   Sydney   Tokyo

ACADEMIC PRESS, INC.
Orlando, Florida 32887

United Kingdom Edition published by
ACADEMIC PRESS, INC. (LONDON) LTD.
24/28 Oval Road, London NW1 7DX

Library of Congress Cataloging in Publication Data

Schowengerdt, Robert A.
    Techniques for image processing and classification
in remote sensing.

    Includes index.
    1. Image processing. 2. Remote sensing. I. Title.
TA1632.S3 1983   621.36'78   83-11769
ISBN 0-12-628980-8 (alk. paper)

PRINTED IN THE UNITED STATES OF AMERICA

85 86 87 88   9 8 7 6 5 4 3 2

To Andrea and Jennifer

Contents

List of Illustrations

Figures

Chapter 1

Plates

Preface

Remote sensing of the earth's surface began with the use of aerial photography in the early 1900s. Aerial mapping cameras and photointerpretation were the tools used until the late 1960s, when the first multispectral scanner systems were flown on aircraft. A parallel interest developed at that time in the quantitative processing and analysis of numerical data from these scanners, and since the advent of the Landsat series of satellites in 1972, digital image processing and classification has become an increasingly important aspect of remote sensing. It is now essential that all students of remote sensing, whether in geology, geography, ecology, or any of the other professions that use remote sensing, be familiar with the techniques used for processing digital images.

This book is designed to introduce these students to computer image processing and classification (commonly called "pattern recognition" in other applications). There has been a need for such a book for some time. Most image processing texts are directed to engineering students and are consequently predominantly mathematical in nature. Furthermore, they usually treat remote sensing as only one application of computer image analysis (justifiably, from their perspective) and include many topics, such as data compression and computer vision, that are of little interest to most remote sensing applications scientists. On the other hand, remote sensing textbooks often present digital image processing in a superficial way, with minimal discussion of the computer algorithms involved, the options that may be available for certain kinds of processing, and the advantages and disadvantages of different processing algorithms. I believe that it is important for scientists who apply remote sensing techniques in their discipline to have a more knowledgeable perspective on digital image processing, without, however, an undue emphasis on the theory and mathematical formulations that may be involved.

I intend this book to be used as either a primary source in an introductory image processing course or as a supplementary text in an intermediate-level remote sensing course. The academic level addressed is upper-division undergraduate or beginning graduate, and familiarity with calculus and basic vector and matrix concepts is assumed. A discussion of digital scanners and imagery and two key mathematical concepts for image processing and classification, spatial filtering and statistical pattern recognition, respectively, are presented in Chapter 1. All or parts of this material may be used at the discretion of the instructor, depending on the desired emphasis in a particular course. Chapters 2 and 3 contain a comprehensive descriptive survey of image processing and classification techniques that are widely used in the remote sensing community. The emphasis throughout is on techniques that assist in the analysis of images, not particular applications of these techniques. A specific effort has been made in the design of new illustrations to enhance teaching of the material; examples drawn directly from the research literature are not always optimal for this purpose. I anticipate, however, that many instructors will add their own or other's research examples to emphasize particular applications or aspects of processing. In this sense, the material in Chapters 2 and 3 can serve as a structured guide to the topics that should be covered in an introductory image processing course with a remote sensing emphasis.

Finally, there are four appendixes, containing a bibliography (independent of the references at the end of each chapter), an introduction to computer binary data representation and image data formats, a discussion of interactive image processing, and a selection of exam questions from the Image Processing Laboratory course at the University of Arizona. Individual instructors may decide to integrate Appendixes B and C on data formats and interactive processing into the main lecture schedule. For example, the subject of interactive image processing can be made an important part of a course if the appropriate computer hardware is available.

As with many textbooks, much of this material originated as notes for teaching, in this case a laboratory course in image processing. This one-semester course is for first- or second-year graduate students and has been attended by electrical and optical engineers, soil scientists, geologists, geographers, computer scientists, and even an occasional astronomer, medical student, or photography major! Conventional classroom lectures are presented from the material in this book, and parallel exercises in batch and interactive image processing using packaged subroutine libraries are conducted by all the students. A computing term project is an option in the second half of the semester, giving the student an opportunity to delve deeper into a particular type of processing or to experiment with image processing in a particular application. A great deal of useful software has originated from these term projects.

In addition to the college environment just described, I anticipate that this book will be useful to those involved in postgraduate remote sensing training, an activity that is especially important in remote sensing because of the relatively short history of formal college programs, both in the United States and elsewhere. Remote sensing and image processing, therefore, are frequently studied as technological subjects by persons already possessing a degree in a traditional earth science discipline.

I am grateful to many people who contributed in a variety of ways to the realization of this book. Primary acknowledgment goes to Karl Glass (Mining and Geological Engineering), Bobby Hunt (Digital Image Analysis Lab), Jack Johnson (Office of Arid Lands Studies), and Phil Slater (Committee on Remote Sensing) for providing a professional environment that was conducive to writing and publishing, not only this book, but also the research results that enhance it. A. P. Colvocoresses and Bob McEwen of the United States Geological Survey also deserve my thanks for giving me the opportunity to learn image processing first hand at the Jet Propulsion Lab and the U.S.G.S. Center for Astrogeology. I much appreciate the first-draft review comments received from Chuck

Hutchinson (Arizona Remote Sensing Center), Dave Nichols (JPL), Steve Park (NASA Langley Research Center), and Gary Peterson (Penn State University). Virtually all of their suggestions are incorporated in one form or another in the final book. I also thank Dick Blackwell (JPL), Ray Jackson (U.S. Department of Agriculture), and Phil Slater for their valuable assistance on specific topics. The excellent artwork was done by Don Cowen (Optical Sciences Center) and Paul Mirocha (Office of Arid Lands Studies), and Anna Elias-Cesnik (Office of Arid Lands Studies) was most helpful in editing the text. All of the word processing was performed by Mike Porter (Mining and Geological Engineering), without whose skill and patience, through numerous rewrites, this book would not have materialized. Finally, I thank my parents for their lifelong encouragement and support and my wife, Amy, for her generous understanding during the writing of this book.

Fundamentals

1.1 Introduction

In this book we discuss the basics of numerical manipulation of remote sensing digital image data. These data commonly come directly from a multispectral scanner system (MSS), such as that on the Landsat satellites, but also may be derived from photographs that have been digitized in the laboratory by an optical scanner. The word "image" acquires a rather general meaning in this context. An image is no longer simply the familiar photographic print or transparency, but is also a two-dimensional array of numbers, each representing the brightness of a small elemental area in the digital image. This numerical representation of images permits the application of a wide assortment of computer processing and analysis techniques to the data. The results of this computer processing are new arrays of numbers, representing improved (enhanced) images or thematic classifications,[1] which then must be converted to an analog representation for display. Figure 1-1 depicts the stages in the numerical processing of remote sensing image data.

In this chapter a conceptual and mathematical framework is presented for understanding digital image processing and classification. The mathematical tools described are particularly

---

[1]For convenience, we will call this type of image-derived product a *map*, although it represents only one component of the many that contribute to a map in the conventional sense of the word.

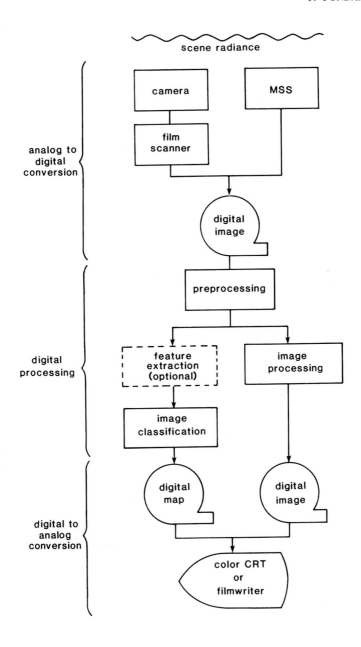

FIGURE 1-1.   *Data flow of digital remote sensing imagery.*

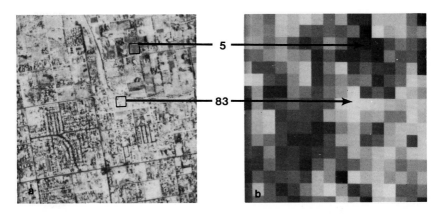

FIGURE 1-2.   Creation of a digital image. (a) Scene. (b) Digital image.

relevant to image manipulation and are necessary for a thorough understanding of much of the material in Chapters 2 and 3, which contain a survey of specific processing techniques that are widely used in the remote sensing community. Additional background in calculus, Fourier theory, and statistics is available from any of the numerous textbooks on these subjects, some of which are listed in the references.

1.2 The Characteristics of Digital Images

Digital images consist of discrete picture elements, called *pixels*. Associated with each pixel is a number that is the average radiance[1] ("brightness") of a relatively small area within a scene, as shown in Fig. 1-2. The size of this area

---

[1]For a good discussion of radiometry and the associated terminology see Slater (1980).

affects the reproduction of detail within the scene, as seen in
Fig. 1-3. As the pixel area is reduced, more scene detail is
preserved in the digital representation. Just as aerial photographs may be optically reduced or enlarged, digital images can
be displayed at any desired scale by appropriate computer
processing (Sec. 2.6). The pixel size of the display device can
also be used to control the final scale of the displayed image.

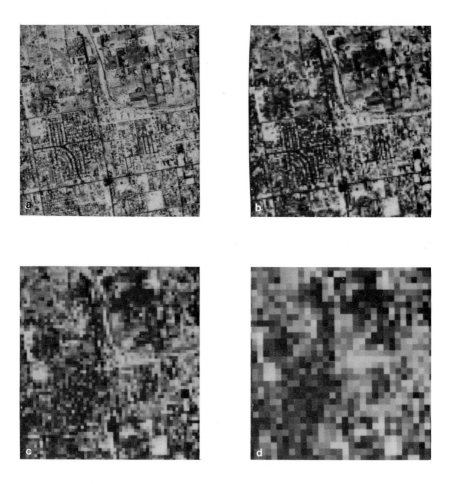

FIGURE 1-3. Digital image structure as a function of pixel size.
(a) 16 m. (b) 32 m. (c) 64 m. (d) 128 m.

For high quality image presentation, the display pixel size is usually small (less than 0.1 mm), so that individual pixels cannot be distinguished at normal viewing distances. For classification maps, however, it often is desirable to use larger display pixels to permit visual examination of the data at the pixel level.

In satellite remote sensing, we are looking through the atmosphere at the earth's surface. The sensor measures not only radiation *reflected* from the surface and transmitted by the atmosphere, but also radiation that is *scattered* by the atmosphere. The value of each pixel in a satellite image of the earth represents the total amount of radiation reaching the sensor and transmitted by the sensor's optics. Fortunately, atmospheric effects are nearly constant over large areas and the *changes* in radiance that the sensor detects are due to changes in the radiance of the ground. The image is thus a useful representation of the radiance of the ground.

1.2.1 Pixel Parameters

The ground area represented by a pixel is determined by the altitude of the sensor system and its design parameters, particularly the instantaneous-field-of-view (IFOV). The IFOV is the angle subtended by the geometrical projection of a single detector element to the earth's surface (Fig. 1-4). There are several scanning methods for moving the sensor's IFOV across the ground (Slater, 1980); all result in a mapping of the continuous two-dimensional scene radiance into a large, two-dimensional array of pixels that constitute a digital image of the scene radiance.

The distance between consecutive measurements of the scene radiance within the IFOV is determined by the sensor system sampling rate, and is usually, but not necessarily, equal to the linear dimensions of the IFOV at the ground. This distance is

commonly referred to as the "size of a pixel", but unless the
sampled IFOVs are contiguous, it does not represent the area
averaged by the IFOV.   An example of such a situation is the
Landsat MSS with overlapping IFOVs in the along-scan (across-
track) direction (Table 1-1).   This overlap results in better
reproduction of scene detail; however, it must be compensated
for to achieve correct geometry if the image is displayed with
equal sample intervals in both directions.

   As with all digital data, a finite number of bits are used
to represent the scene radiance for each pixel.   The continuous
radiance of the scene is therefore *quantized* into discrete *gray
levels* in the digital image.   Only about 5 or 6 bits per pixel

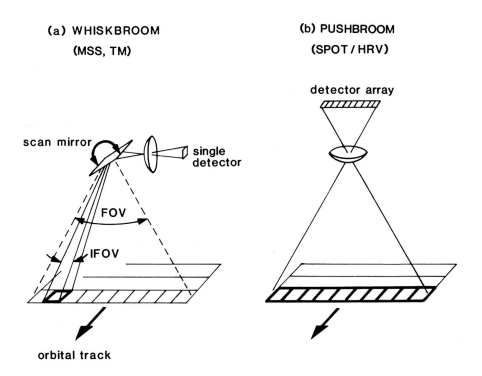

(a) WHISKBROOM
(MSS, TM)

(b) PUSHBROOM
(SPOT / HRV)

detector array

scan mirror
single detector

FOV

IFOV

orbital track

FIGURE 1-4.   Sensor scanning methods.

TABLE 1-1. Characteristics of Several Remote Sensing Systems[1]

	Landsat MSS	Thematic Mapper (TM)	SPOT/HRV
Launch	1 1972 2 1975 3 1978 4 1982	1982	1984
Altitude (km)	920 (1–3) 695 (4)	695	822
Spectral bands (µm)	4 0.5–0.6 5 0.6–0.7 6_2 0.7–0.8 7^2 0.8–1.1	1 0.45–0.53 2 0.52–0.60 3 0.63–0.69 4 0.76–0.90 5 1.55–1.75 6 10.40–12.50 7 2.08–2.35	1 0.50–0.59 2 0.61–0.68 3 0.79–0.89 p 0.51–0.73[3]
IFOV (m)	76 x 76 (1–3) 80 x 80 (4)	30 x 30 (bands 1–5,7) 120 x 120 (band 6)	20 (bands 1–3) 10 (p)
Pixel interval (m)	57 x 82 (1–3) 57 x 80 (4)	30 x 30 (bands 1–5,7) 120 x 120 (band 6)	20 (bands 1–3) 10 (p)
FOV (km)	185 x 185	185 x 185	60 x 60[4]
Pixels/ scene ($\times 10^6$)	28	231	27 (bands 1–3) 36 (p)
Bits/pixel	6	8	8 (bands 1–3) 6 (p)

[1]Compiled in part from Slater (1980). Values for Landsat-4 MSS, TM, and SPOT/HRV are nominal design values.

[2]The MSS bands on Landsat-1, -2, and -3 were numbered 4, 5, 6, and 7 because of a three-band return beam vidicon (RBV) sensor on -1 and -2. Beginning with Landsat-4 the MSS bands are renumbered 1, 2, 3, and 4.

[3]panchromatic mode.

[4]The SPOT satellite will carry two sensors that are pointable across the orbital track. A 120-km total FOV coverage will therefore be possible.

(32 or 64 gray levels, respectively) are required to yield a visually continuous range of brightness in a displayed digital image. More bits per pixel are desirable for numerical analyses of the data, however. The Landsat MSS has 6 bits per pixel, but the next generation sensor systems will have 8, resulting in 256 gray levels (Table 1-1). The effect of this parameter on the visual appearance of an image is shown in Fig. 1-5. As the number of gray levels is reduced, the image becomes mottled and spatial detail is lost.

In summary, a pixel is completely characterized by three quantities: 1) the linear dimension of the sensor IFOV projected to the ground; 2) the distance between consecutive IFOV samples; and 3) the number of bits representing the measured radiance. These parameters specify the *ideal* spatial and radiometric *resolution* of the final digital image. Many other factors affect the *actual* resolution of the image, however. For example, spatial resolution depends not only on the IFOV and the distance between IFOV samples, but also on IFOV motion that occurs during the pixel sampling time because of scanning or platform motion, and the electronic characteristics of the sensor and data transmission equipment. Furthermore, as seen in Figs. 1-3 and 1-5, spatial and radiometric resolution can interact in determining the overall quality of an image.

The word pixel is used rather freely in the remainder of this book (as it is in the remote sensing community) to describe one of the numbers that constitute a digital image, but the various physical quantities implied in that usage should always be kept in mind.

1.2.2 Image Parameters

The field-of-view (FOV) of the sensor (Fig. 1-4) determines the ground area covered by an image and, coupled with the distance between IFOV samples, determines the total number of pixels in the image. In remote sensing the number of pixels per image is quite large, on the order of tens of millions (Table 1-1), and consequently affects every aspect of image acquisition, processing, display and storage. Only the continuing improvements in digital electronics and computer hardware

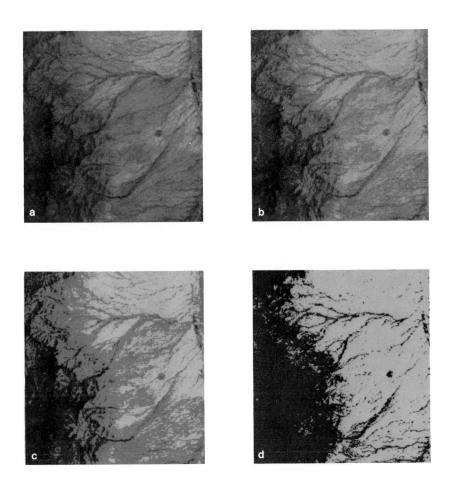

FIGURE 1-5. Digital image structure as a function of number of gray levels. (a) 16. (b) 8. (c) 4. (d) 2.

and software have made possible the routine processing of such
large amounts of data.

Remote sensing images are commonly *multispectral*, i.e., the
same scene is imaged simultaneously in several spectral bands of
the electromagnetic spectrum. An example of a multispectral
image in the visible and near infrared portions of the spectrum
is the Landsat MSS image in Fig. 1-6. The four images are
registered so that each pixel actually has associated with it
four gray levels, one in each spectral band. The radiance
measured in each band is an average value over a fairly broad
spectral band (Table 1-1; Fig. 3-3). The amount of image data
generated by a given sensor is directly proportional to the
number of spectral bands if they have equal IFOVs. It has been
shown, however, that because of band-to-band spectral
correlation, it is not necessary to design each band with
equally high spatial resolution (Schowengerdt, 1980).

Image *contrast* is related to the range of gray levels in an
image; the greater the range, the greater the contrast and vice
versa. Contrast, C, may be defined numerically in several ways,
e.g.

$$C_1 = GL_{max}/GL_{min}$$

or $\quad C_2 = GL_{max} - GL_{min}$

or $\quad C_3 = \sigma_{GL}$ \hfill (1-1)

where GL_{max} and GL_{min} are the maximum and minimum gray levels in
the image, and σ_{GL} is the gray level standard deviation. Each
of these definitions has advantages and disadvantages in
particular applications. For example, one or two bad pixels in
a large image could result in deceptively high values for C_1 and
C_2, whereas C_3 would be much less affected.

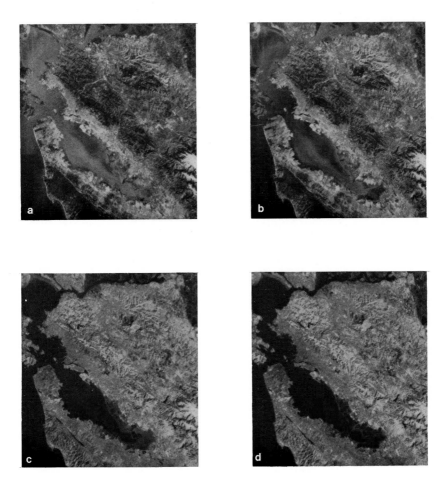

FIGURE 1-6. A Landsat MSS image of San Francisco, California.
(a) Band 4. (b) Band 5. (c) Band 6. (d) Band 7.

Contrast is an important image property for two reasons. First, the numerical definitions of Eq. (1-1) may be used to measure the *signal-to-noise ratio* (SNR) of the digital image data. For example, for an image contaminated by uniform random noise (see Sec. 2.4) the SNR often is defined as the ratio of

the noise-free image contrast to the noise contrast, represented
in both cases by either C_3 or $C_3{}^2$. Second, the contrast of a
displayed image is one indicator of the visual quality of the
image. In this context, the quantities in Eq. (1-1) should not
be gray levels in the digital image, but rather direct measurements of the radiance of the displayed image.

The *visually perceived* contrast of a displayed image
depends not only on its gray level range, but also on psycho-
physical factors such as the spatial structure within the image
(see Cornsweet, 1970, for examples) and the ambient light level
in the viewing area. Furthermore, both visual and numerical
contrast are area-dependent quantities in that relatively small
regions of an overall high contrast image may have low, high, or
intermediate contrast. Local contrast enhancement (Sec. 2.2) is
a processing technique designed to partially remove this
regional variation.

Another measurable image quantity, related to contrast, is
modulation, M, defined as

$$M = \frac{GL_{max} - GL_{min}}{GL_{max} + GL_{min}} \qquad (1-2)$$

Because gray levels are always positive, the definition in
Eq. (1-2) insures that modulation is always between zero and
one. Modulation is most appropriately used to describe periodic
(repetitive) signals but is sometimes used for nonperiodic
signals as well. Note that modulation is related to one measure
of contrast, C_1, as follows:

$$M = \frac{C_1 - 1}{C_1 + 1} \qquad (1-3)$$

The modulation transfer function (MTF) discussed in Sec. 1.4 describes how an optical system reduces scene modulation in the imaging process, or how computer processing can be used to decrease or increase the modulation in digital images.

Although digital images may be displayed with small pixels, such that they appear continuous, it is important to keep in mind their *discrete* spatial and radiometric nature, which can cause artifacts. For example, if there are periodic patterns in the scene, such as agricultural crops planted in rows, and their spatial period is about the same as the pixel sample interval, an interference pattern can result as seen in Fig. 1-7a. This phenomenon, known as *aliasing*, has not been commonly observed in satellite imagery because large, high contrast periodic ground patterns occur infrequently. The higher ground resolution of new sensors may increase the likelihood of this problem, however. Other scanner artifacts are much more common, such as the "stairstep" appearance of linear features (Fig. 1-7b).

1.3 The Distinction Between Image Processing and Classification

Digital image processing is the numerical manipulation of digital images and includes *preprocessing, enhancement* and *classification*. Preprocessing refers to the initial processing of the raw data to calibrate the image radiometry, correct geometric distortions, and remove noise. The nature of the particular preprocessing required obviously depends strongly on the sensor's characteristics, because the preprocessing is designed to remove any undesirable image characteristics produced by the sensor. The corrected images are then submitted to enhancement or classification processing, or both.

Image enhancement produces a new, *enhanced* image that is displayed on a cathode ray tube (CRT), for example, for visual interpretation. This enhanced image may be easier to interpret

FIGURE 1-7. *Scanner artifacts. (a-c) Periodic patterns (Legault, 1973). (d, e) Linear patterns: d, scene; e, scanner image (Biberman, 1973).*

than the original image in different ways. For example, more efficient use may be made of the original information (contrast enhancement), or additional visual dimensions may be used to emphasize subtle information (color enhancement). It is not necessary that the enhanced image look like a conventional image, but the changes that have been caused by processing should be understood to permit correct visual interpretation.

Image classification carries the digital processing a step further and attempts to replace the visual interpretation step with quantitative decision making. The output from classification processing, therefore, is a *thematic map*, in which each pixel in the original imagery has been classified into one of several "themes," or classes. Although the intent is to make the mapping process more quantitative and objective, human input and interaction with the processing is a vital part of a successful classification. Classification plays virtually no role in some remote sensing applications, such as geologic lineament mapping, but image enhancement can be extremely useful. The two types of processing thus complement each other and the decision to employ one, or perhaps both, in a particular application can be made only with an understanding of the characteristics of these two approaches to the extraction of information from images.

Although preprocessing, enhancement and classification may be considered distinct topics as just discussed, there are many interrelationships in practice. For example, some preprocessing techniques, such as noise suppression (Sec. 2.4), may just as well be considered image enhancement techniques. Some processing techniques, such as spectral band ratios (Sec. 3.4.2), have been useful for both producing enhanced images and improving classifications. The latter is one example of how classification accuracy often is improved by judicious pre- or post-classification processing (Secs. 3.4 and 3.6). The curious

mixture of terms "classification enhancement" might be applied
to this type of processing!

Because of the functional similarity of many of the
techniques used for image preprocessing and image enhancement,
we will simply use the term "image processing" to include both
preprocessing and enhancement. The term "image classification"
will include those techniques that are primarily used to produce
thematic maps from images.

1.4 Mathematical Concepts for Image Processing

The wide assortment of image processing techniques
described in Chapter 2 may be divided into two generic
types of processing (Fig. 1-8). *Point* processing, the simplest
type, consists of a transformation of each original image pixel
value into a new value for the output image. The transformation
depends only on the gray level of the original image pixel and
includes techniques such as contrast enhancement and multispectral ratios.

Neighborhood processing also performs a transformation on
each pixel in a way that depends not only on the gray level of
the pixel being processed but also on the gray levels of pixels
in the vicinity of the pixel being processed and includes techniques such as edge enhancement and interpolation. If this
localized influence may be expressed by a weighted sum of pixels
in the neighborhood of the pixel being processed, a process
known as *linear spatial filtering* results. There is a strong
analogy between linear spatial filtering and the formation of
images by an optical system, a connection that is emphasized as
a conceptual aid in the following discussion. For convenience
in notation, the mathematical treatment is primarily in terms of
continuous functions; a discussion of the discrete implementation of spatial filtering, and a related topic, correlation, is

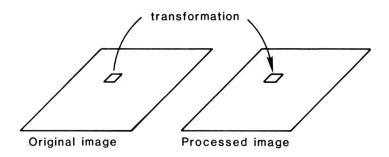

(a) PIXEL TRANSFORMATION

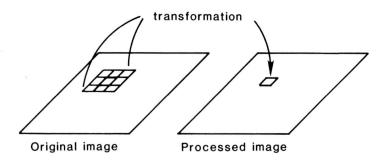

(b) NEIGHBORHOOD TRANSFORMATION

FIGURE 1-8. The two general types of image processing.

provided in Sec. 1.4.5. Because of its relative simplicity, point processing does not require a mathematical discussion here.

1.4.1 Image Formation

An image forming system may be treated as a "black box" that operates on an input signal to produce an output signal (Gaskill, 1978). The input signal is the scene radiance and the output signal is the image irradiance; both are two-dimensional

functions of spatial coordinates. Figure 1-9a depicts such a
system at 1:1 magnification and in one dimension, for convenience. We are not concerned here with the components of this
system and therefore simply represent its effect on the input
signal by an operator $\mathscr{L}$. If the input to this system consists
of a single impulse (Fig. 1-9b), the output is termed the
impulse response. In optical terminology the impulse response
is called the *point spread function* (PSF) and is the two-
dimensional image of a point source. The size and shape of the
PSF is a measure of the system's imaging performance and is
determined by the optical system's F-number, aberrations, the
wavelength used, and other, external factors such as atmospheric
turbulence and sensor vibration (Slater, 1980). Generally, the
narrower the PSF, the better the system and the imagery it
produces.

If the input signal consists of two or more impulses, and
the output signal is the sum of the outputs produced by each
impulse (Fig. 1-9c), the system is termed *linear*. Furthermore,
if a spatial shift of the input signal produces a corresponding
shift in the output, but otherwise no change in the PSF, the
system is *shift-invariant* (Fig. 1-9d). We may summarize these
conditions mathematically in the following way,

system description $\quad g(x,y) = \mathscr{L}[f(x,y)]$

linear system $\qquad\quad g(x,y) = \mathscr{L}[f_1(x,y) + f_2(x,y)]$

$$= \mathscr{L}[f_1(x,y)] + \mathscr{L}[f_2(x,y)]$$

$$= g_1(x,y) + g_2(x,y)$$

and

shift-invariant
system $\qquad\quad g(x-x',y-y') = \mathscr{L}[f(x-x',y-y')] \qquad (1-4)$

where $f(x,y)$ is the scene radiance and $g(x,y)$ is the image
irradiance.

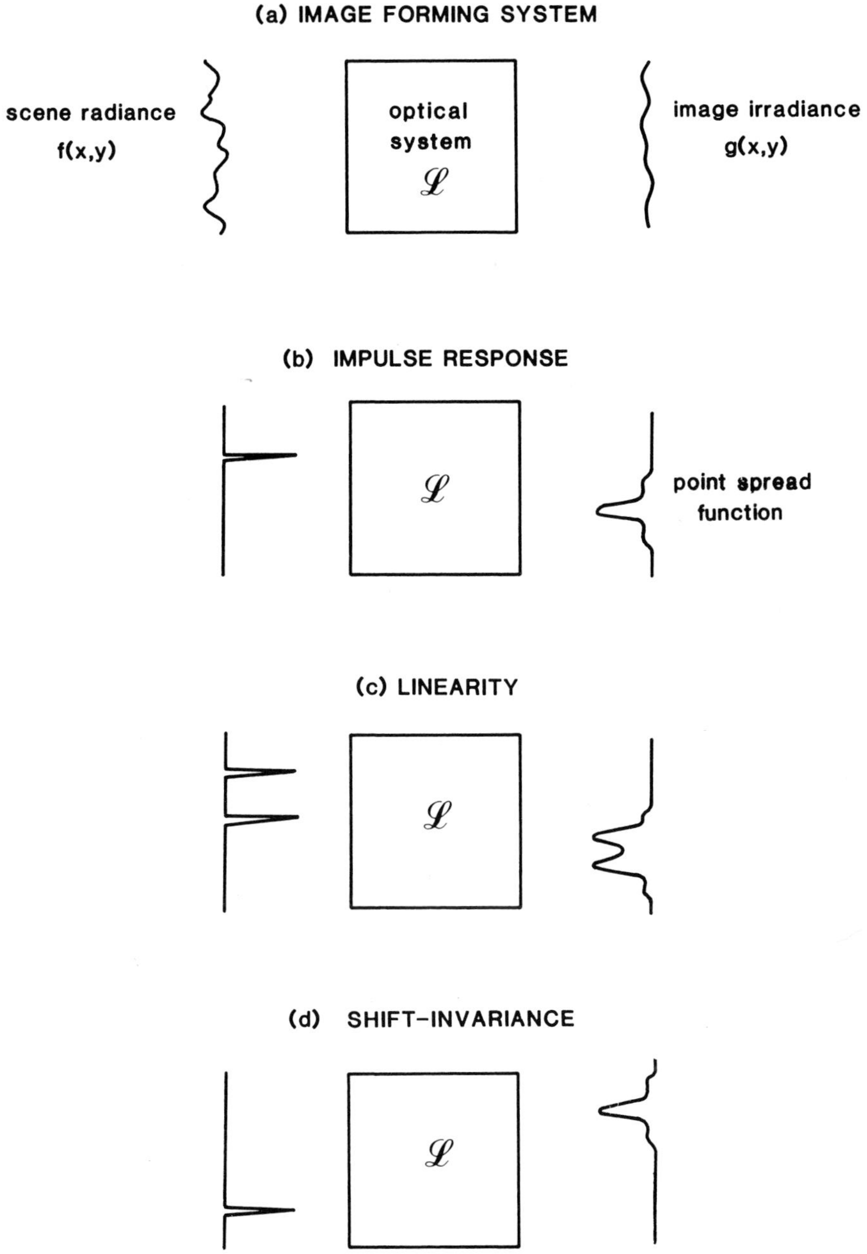

(a) IMAGE FORMING SYSTEM

scene radiance
f(x,y)

optical
system
$\mathscr{L}$

image irradiance
g(x,y)

(b) IMPULSE RESPONSE

$\mathscr{L}$

point spread
function

(c) LINEARITY

$\mathscr{L}$

(d) SHIFT—INVARIANCE

$\mathscr{L}$

FIGURE 1-9. *Description of an optical system as a linear system.*

If a system is linear and shift-invariant (LSI), the operator, $\mathscr{L}$, may be described, for any input signal, as a *convolution* of the PSF with the input signal. This is commonly represented by the notation

$$g(x,y) = PSF(x,y)*f(x,y) \qquad (1-5)$$

where the * symbol indicates the convolution operation given by

$$PSF(x,y)*f(x,y) = \int\int_{-\infty}^{+\infty} PSF(x',y')f(x - x',y - y')dx'dy'$$

$$= \int\int_{-\infty}^{+\infty} f(x',y')PSF(x - x',y - y')dx'dy' \qquad (1-6)$$

The last equality means that convolution is commutative, i.e., f*PSF equals PSF*f. Equation (1-6) is the fundamental equation for linear spatial filtering. The value of the output image at any point (x,y) is given by a weighted summation of the input signal, the scene radiance, in the vicinity of (x,y). The weighting is determined by the PSF. An optical PSF can be emulated in the processing of digital images by using an appropriate weighting on the pixels surrounding the pixel being processed (Sec. 1.4.5).

Figure 1-10 depicts the convolution of Eq. (1-6) in one dimension. The PSF is first inverted as a function of the integration variable x' and is then shifted by an amount x along the x' axis. The value of the convolution g(x) is given by the area under the product of f(x) and the inverted, shifted PSF(x). For well-designed optical systems, or in digital image processing, most PSFs are symmetric and the inversion operation has no effect. In the few cases where the PSF is asymmetric, such as a

one-dimensional digital derivative (Fig. 2-14), inversion of the
PSF is irrelevant in terms of the utility of the enhanced image.

As indicated in Fig. 1-10, the image produced by the convolution between the scene radiance and the optical PSF is a
smoothed representation of the scene. In other words, the scene
radiance modulation is *reduced* by the imaging process. This
reduction of modulation is characteristic of all imaging and
scanning systems and results from the all-positive nature and
non-zero width of optical PSFs. For *digital* spatial filtering,
we have considerable flexibility in defining a PSF and often use
negative weights in the PSF to increase the image modulation.
The visual result of such processing is a "sharpening" of fine

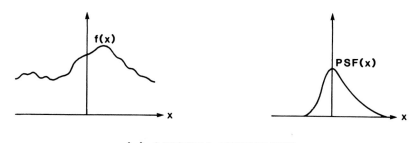

(a) ORIGINAL FUNCTIONS

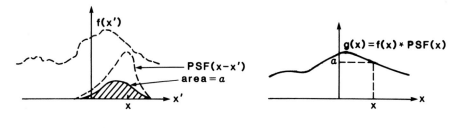

(b) SHIFT-MULTIPLY-INTEGRATE   **(c) RESULT**

FIGURE 1-10. Convolution of two functions.

detail in the image.   Chapter 2 contains several examples of this and other types of digital spatial filtering.

1.4.2 The Fourier Transform Approach

Thus far, the discussion of linear spatial filtering has been in terms of functions of the *spatial domain*, i.e., the (x,y) coordinate space of images.   Another coordinate space, defined by the Fourier transform and known as the spatial *frequency* domain, is useful in the analysis and understanding of spatial filtering.

It is well known that a periodic function may be represented by an infinite, weighted sum of trigonometric sine and cosine functions with different amplitudes, frequencies and phases.   This representation of a function is termed its Fourier series.   For nonperiodic functions, the infinite series becomes an infinite integral of the form

$$F(\nu_x) = \int_{-\infty}^{+\infty} f(x)e^{-i2\pi\nu_x x}dx \qquad (1-7)$$

where $\quad e^{-i2\pi\nu_x x} = \cos(2\pi\nu_x x) - i\ \sin(2\pi\nu_x x)$

and i is an "imaginary number" equal to $\sqrt{-1}$ (Churchill and Brown, 1978).

Equation (1-7) defines the one-dimensional *Fourier transform*, $F(\nu_x)$, of f(x). $F(\nu_x)$, called the *spatial spectrum* of the spatial function f(x), is a complex function (Churchill and Brown, 1974; Bracewell, 1978) of *spatial frequency* ν_x with the real part

$$Re[F] = \int_{-\infty}^{+\infty} f(x)\cos(2\pi\nu_x x)dx \qquad (1-8)$$

and imaginary part

$$\text{Im}[F] = -\int_{-\infty}^{+\infty} f(x)\sin(2\pi\nu_x x)dx \qquad (1\text{-}9)$$

The *amplitude*, or modulus, of F is given by

$$\begin{aligned} \text{Am}[F] &= \sqrt{\text{Re}^2[F] + \text{Im}^2[F]} \\ &= \sqrt{FF^*} \end{aligned} \qquad (1\text{-}10)$$

and the *phase* is given by

$$\text{Ph}[F] = \tan^{-1}(\text{Im}[F]/\text{Re}[F]) \qquad (1\text{-}11)$$

where F^* is the complex conjugate of F obtained by changing the sign of the imaginary part, Eq. (1-9).

In two dimensions Eq. (1-7) is written as

$$F(\nu_x,\nu_y) = \int\!\!\!\int_{-\infty}^{+\infty} f(x,y)e^{-i2\pi(\nu_x x + \nu_y y)}dxdy \qquad (1\text{-}12)$$

For example, if $f(x,y)$ is an image, then $F(\nu_x,\nu_y)$ is the image spatial spectrum as a function of the spatial frequency coordinates, ν_x and ν_y. The units of spatial frequency are cycles/unit length, e.g., cycles/mm at image scales or cycles/km at the ground.

It can be shown (Bracewell, 1978) that the *inverse* Fourier transform is obtained by simply interchanging $f(x,y)$ and $F(\nu_x,\nu_y)$, changing the variable of integration, and changing the sign of the exponent in Eq. (1-12),

$$f(x,y) = \int\!\!\!\int_{-\infty}^{+\infty} F(\nu_x,\nu_y)e^{+i2\pi(\nu_x x + \nu_y y)}d\nu_x d\nu_y \qquad (1\text{-}13)$$

In the inverse Fourier transform, $F(\nu_x, \nu_y)$ weights the cosine and sine components of $f(x,y)$. The amplitude contribution to $f(x,y)$ of a pair of cosine-sine components with a given frequency is given by $Am(F)$. That is, $Am[F(\nu_x, \nu_y)]$ is the "strength" of the components with spatial frequency (ν_x, ν_y). The superposition of two-dimensional sinusoidal functions in the spatial domain and their corresponding Fourier transforms are shown in Fig. 1-11. Note that the amplitude of $F(\nu_x, \nu_y)$ is proportional to the modulation of $f(x,y)$ [Eq. (1-2)] for single frequency components. In general, the higher frequency components of $F(\nu_x, \nu_y)$ contribute to the *sharpness* of edges in the image, and the lower frequency components contribute to the overall *contrast* of the image.

The greatest utility of the Fourier transform is its application to the linear filtering operation, Eq. (1-5). By a familiar property of Fourier transforms, known as the *convolution theorem* (Bracewell, 1978), Eq. (1-5) can be written

$$G(\nu_x, \nu_y) = OTF(\nu_x, \nu_y)F(\nu_x, \nu_y) \qquad (1-14)$$

where G, F and OTF are the Fourier transforms of g, f and PSF, respectively. G is the image spatial spectrum, F is the scene spatial spectrum and OTF is the *Optical Transfer Function*. The most important aspect of Eq. (1-14) is that, with the use of the Fourier transform, the cumbersome integration in Eq. (1-6) has been replaced by a simple multiplication of functions in the Fourier domain. We can return to the spatial domain if desired by applying the inverse Fourier transform, Eq. (1-13), to $G(\nu_x, \nu_y)$ to calculate the image, $g(x,y)$.

1.4.3 The Optical Transfer Function

Equation (1-14) is not only mathematically simple compared to Eq. (1-5), but also provides an alternative visualization of

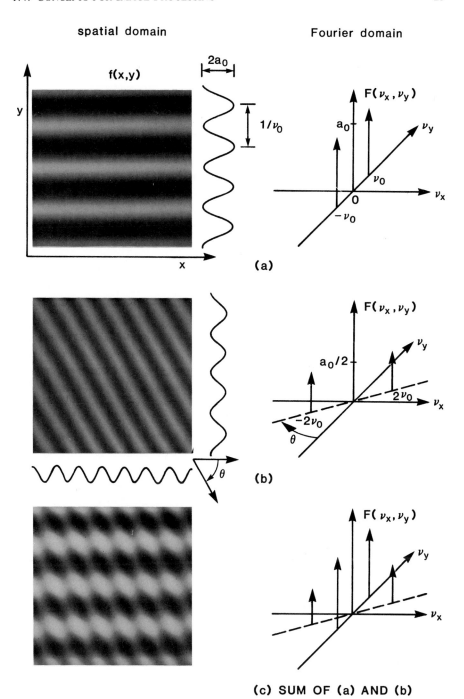

FIGURE 1-11. *Spatial and frequency domain representation of two-dimensional sinusoidal functions.*

spatial filtering. The OTF is the optical analogy to a *low-pass* electronic signal filter, i.e., it attenuates the amplitudes of the high spatial frequency components of the scene spectrum, thus producing an image that is a blurred version of the original scene. The spatial frequency value labeled ν_c in Fig. 1-12 is called the system *cutoff frequency*, and, like the width of the PSF, is a fundamental measure of the optical system's imaging performance.   Figure 1-12 illustrates how ν_c affects image sharpness.

The *modulation transfer function* (MTF) is referred to frequently in the literature.   The MTF is the modulus of the complex function OTF. Using our earlier notation,

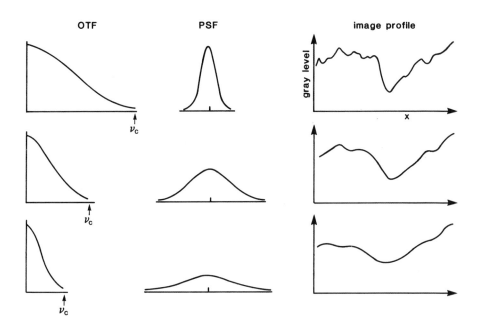

FIGURE 1-12.   *The effect of the cutoff frequency* ν_c *of a low-pass spatial filter.*

$$MTF = Am[OTF] \qquad (1-15)$$

For well-designed optical systems the PSF is symmetric, making the imaginary part of the OTF identically zero. Under these conditions, the MTF is simply the absolute value of the real part of the OTF (see Eqs. [1-10] and [1-11]). The MTF is a sufficient descriptor of an optical system in many cases; it is wise to remember, however, that the full, complex OTF must sometimes be used, particularly if the PSF is asymmetric.

A graphical representation of image formation in the spatial and Fourier domains is shown in Fig. 1-13. The scene consists of rectangular areas, each with a different radiance level, that could represent, for example, agricultural fields. The PSF causes considerable blurring in the image. Our ability to detect the individual rectangles of the original scene in the blurred image depends on their size and the contrast between adjacent rectangles. The severe degradation evident in the spatial domain is explained in the Fourier domain by the low cutoff frequency of the OTF, which strongly attenuates the high spatial frequency components in the scene spectrum.

1.4.4 Image Correlation

Remote sensing projects frequently require accurately registered images from different dates or sensors. A technique commonly employed to register two images is visual location of *control points* (landmarks) that are common to both images, followed by a geometric transformation of one image to match the other (Sections 2.5 and 2.6). There is a great deal of interest in automating the registration of control points, particularly for automatic compilation of topographic maps from large quantities of stereo imagery. The automatic registration of control points between images is known as *correlation*. Although not

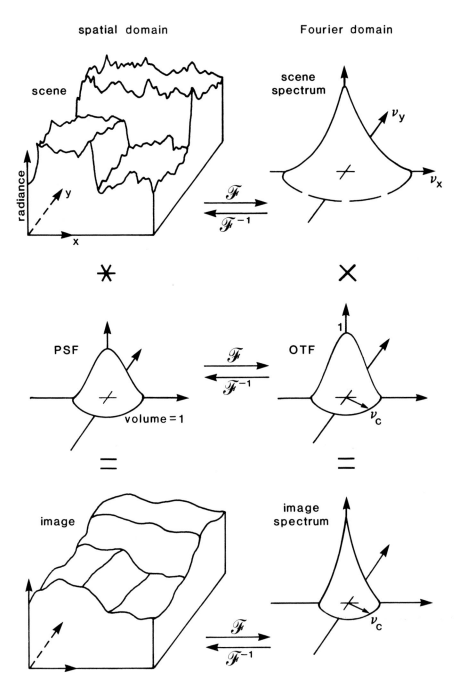

FIGURE 1-13. *Image formation in the spatial and Fourier domains.*

directly related to image formation or spatial filtering, image correlation is mathematically similar and therefore warrants a brief discussion at this point.

All digital correlation algorithms rely on some similarity criterion between the two images to be registered. One such criterion is the area under the product of the two images, as a function of relative spatial shift between them. This particular definition of the correlation, $r(x,y)$, between two similar images, $f_1(x,y)$ and $f_2(x,y)$, is given by

$$r(x,y) = f_1(x,y) \star f_2(x,y)$$

$$= \int\int_{-\infty}^{+\infty} f_1(x',y')f_2(x + x',y + y')dx'dy' \qquad (1-16)$$

Note the similarity between Eqs. (1-16) and (1-6). The only difference is that neither f_1 or f_2 is inverted in the correlation, which is only reasonable since we are looking for the value of (x,y), i.e., the relative spatial shift, where the two images match. Figure 1-14 depicts the correlation of Eq. (1-16) in one dimension. Just as in convolution, a series of "shift-multiply-integrate" operations is performed in correlation.

The Fourier transform of Eq. (1-16) produces a result similar to Eq. (1-14), i.e., the *cross-correlation spectrum* given by

$$R(\nu_x,\nu_y) = F_1(\nu_x,\nu_y)F_2^*(\nu_x,\nu_y) \qquad (1-17)$$

The fact that neither f_1 nor f_2 are inverted in Eq. (1-16) results in the complex conjugate appearing in Eq. (1-17). The

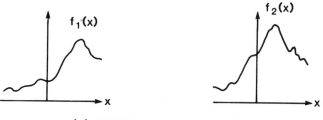

(a) ORIGINAL FUNCTIONS

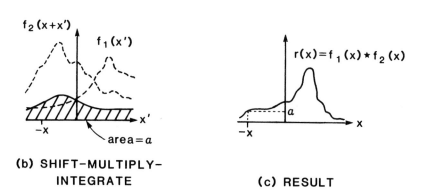

(b) SHIFT-MULTIPLY-
     INTEGRATE   **(c) RESULT**

FIGURE 1-14. Correlation of two functions.

simplicity of Eq. (1-17) compared to Eq. (1-16) again points out the mathematical advantage of the Fourier domain representation. Equation (1-17) has been used in practice for image correlation (Anuta, 1970).

If f_1 and f_2 are sufficiently similar, except for a relative spatial shift, Eq. (1-16) should result in a maximum for $r(x,y)$ at the point of best alignment. Large changes in the image mean level over the region of correlation, however, can lead to false correlation peaks. Therefore, definitions of correlation that are less sensitive to the mean level have been proposed, for example, the total absolute difference

$$r(x,y) = \int\limits_{-\infty}^{+\infty}\int |f_1(x',y') - f_2(x + x',y + y')| dx'dy' \qquad (1\text{-}18)$$

This and other measures of correlation, however, do not possess the linear superposition property of Eq. (1-16) and the attendant mathematical advantages of Eq. (1-17). We shall see in the next section, however, that the simplicity of the Fourier domain representation for convolutions and correlations is often only an aid for mathematical analysis, and these operations are most efficiently implemented in the spatial domain on the computer.

1.4.5 Digital Implementation

The implementation of the mathematical operations discussed in the previous sections in terms of discrete image data is straightforward. For discrete images of finite size (N pixels-by-M lines), Eqs. (1-6), (1-16) and (1-12) become[1]

$$g(i,j) = \sum_{m=1}^{M} \sum_{n=1}^{N} f(m,n)\,PSF(i - m, j - n) \qquad (1\text{-}19)$$

$$r(i,j) = \sum_{m=1}^{M} \sum_{n=1}^{N} f_1(m,n)f_2(i + m, j + n) \qquad (1\text{-}20)$$

$$F(k,\ell) = \sum_{m=1}^{M} \sum_{n=1}^{N} f(m,n)e^{-i\,2\pi\,(km/M + \ell n/N)} \qquad (1\text{-}21)$$

---

[1]In these and subsequent equations, we ignore complications in calculation and notation near the border of the image. For a discussion of the "border problem" see Pratt (1978, pp. 288-289).

The indices i, j, m and n are discrete integers corresponding to the line number or pixel number within a line. Note that the Fourier transform has acquired M^{-1} and N^{-1} factors in the exponent which normalize the spatial frequency indices (k,ℓ) to units of cycles/unit length.

Equations (1-19) and (1-20) may be modified to greatly reduce the number of calculations required. In the case of convolution, Eq. (1-19), the PSF usually has a relatively small width compared to the size of the image. That is, the contributions to the output pixel at (i,j) come from a small region about the input pixel at (m,n) = (i,j). If the width of the PSF is W pixels-by-W lines, we can rewrite Eq. (1-19) as

$$g(i,j) = \sum_{m=i-W/2}^{i+W/2} \sum_{n=j-W/2}^{j+W/2} f(m,n) \, PSF(i-m,j-n) \qquad (1-22)$$

The number of calculations required *for each output pixel* is now W^2 compared to MN for Eq. (1-19). We can think of the operation in Eq. (1-22) in terms of a window, W-by-W pixels large, moving through the input array f(m,n). The input pixels within the window are multiplied by the corresponding PSF values (weights) and then summed to create each output pixel. The window then moves over one pixel in the same line and the process is repeated using the original input pixels. This moving window concept is depicted in Fig. 1-15. Generally, W is an odd number for symmetry reasons, i.e., so that the input and output images are exactly registered, but the window does not have to be square. Specific PSF examples are discussed in Sec. 2.3.1.

A similar efficiency may be achieved for Eq. (1-20) using somewhat different arguments. In a correlation, we are looking

PSF

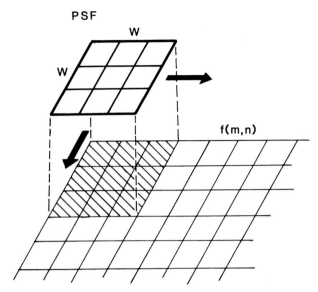

(a) PROJECTION OF WINDOW ONTO IMAGE BEING PROCESSED

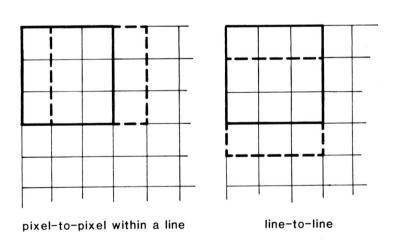

pixel-to-pixel within a line   line-to-line

(b) CONSECUTIVE WINDOW CALCULATIONS

FIGURE 1-15. The moving spatial window concept.

for a local peak (maximum) in r(i,j) indicating where the two
images are correctly registered. It is anticipated that r(i,j)
will decrease fairly rapidly away from this peak as the shift
between the two images increases. Therefore, if the two images
are *approximately registered* before correlation, only small
shifts near (m,n)=(i,j) are needed to find the peak in r(i,j).
Also, the correlation is not normally performed over the total
overlapping area of the two images but only within a relatively
small window (Sec. 2.5.1), which further reduces the computational burden.

For general functions f(m,n) there are no obvious simplifi-
cations for Eq. (1-21). The development of the fast Fourier
transform (FFT) algorithm in 1965 (Brigham, 1974), however,
provided an extremely efficient approach to the calculation.
The number of mathematical operations required by Eq. (1-21) to
calculate $F(k,\ell)$ for each (k,ℓ) is proportional to M^2N^2. This
is more than a *billion* calculations for only a 512^2 image. By
ingeniously reorganizing the data and taking advantage of the
periodicities in the complex exponential term, the FFT can
perform the transform with about $MNlog_2MN$ calculations (Brigham,
1974). For a 512^2 image this is more efficient by a factor of
about 3×10^3.

Although the FFT makes Fourier transforms of images a
feasible proposition, its use has generally been limited to
images no larger than 512^2 pixels for several reasons. First,
the most efficient algorithms require that N and M be powers of
two (128, 256, 512, 1024, etc.). Thus, image arrays of other
sizes must either be "padded" (surrounded) by zero-valued pixels
to increase N and M to the nearest power of two, or the transform must be performed in a cumbersome series of blocks in the
original image. Furthermore, the most efficient FFT algorithms
require that the entire array reside in memory, a requirement
that is prohibitive for images that are 1024^2 pixels or

larger. FFTs that utilize disk storage are considerably less efficient. Finally, if the computation requirements for Eqs. (1-21) and (1-22) are carefully compared, remembering that the inverse Fourier transform also must be calculated in a Fourier domain filtering operation, we find that the spatial domain approach is *always* more efficient if W is less than 8, no matter how large the image is (Pratt, 1978). Therefore, for spatial filtering of the large images encountered in remote sensing, the spatial domain calculation of Eq. (1-22) is used almost exclusively.

1.5 Mathematical Concepts for Image Classification

As discussed earlier in this chapter, there is a fundamental difference between image enhancement and image classification. Image enhancement is designed to enhance the image that is displayed for visual interpretation by the analyst; the decision process required to produce maps from these images remains the task of the photointerpreter. Image classification, on the other hand, assigns the decision-making process to the computer. The intent is to replace the sometimes vague or ambiguous interpretations of the analyst by more quantitative and repeatable processes. Image classification of satellite data by computer has the potential for efficient and consistent mapping of large areas of the earth's surface.

Because image classification is essentially a decision-making process with data that can exhibit considerable statistical variability (Sec. 3.2.2), we must rely on the mathematical tools of statistical decision theory. At best, the decision to classify a pixel into any particular class is a statistically intelligent "guess," which has some associated probability of error. Consequently, it is logical to require that the decision made at each pixel minimize some error criterion throughout the classified area, i.e., over a large number of individual pixel classifications.

An intuitively satisfying and mathematically tractable classification theory having the above property is *maximum-likelihood*, or Bayes optimal, classification. We will review the basic mathematics of this approach to illustrate the concepts behind statistical classification in general. The basic Bayes theory, the resulting rules for making classification decisions, and how the mathematics and functions change in going from one-dimensional to K-dimensional data are described. Finally, a discussion of an important special case of the maximum-likelihood approach, the minimum-distance algorithm, is presented. Practical aspects of implementing classification of remote sensing images are discussed in Chapter 3.

1.5.1 Bayes Theory

Suppose we measure some *feature* of a scene (for example, the gray level of each pixel) and must decide to which of two classes (for example, vegetation or soil) a pixel belongs. This is a one-dimensional, two-class classification problem in the feature domain of the image. If a large number of pixels are available that may be considered representative of each class (i.e., *training data*, Sec. 3.3), we can calculate a *relative frequency* histogram of the feature for each class (Fig. 1-16a) and consider these to be approximations to the continuous probability density functions of an infinite sample of data. These *state-conditional* probability density functions, $p(x|1)$ and $p(x|2)$, have unit area and describe the probability of a pixel having a feature value x *given* that the pixel is in class 1 or class 2, respectively.

Each probability density function (histogram) may be scaled by the *a priori* probability, $p(i)$, that class i occurs in the image area of interest (Fig. 1-16b). These scaled probability

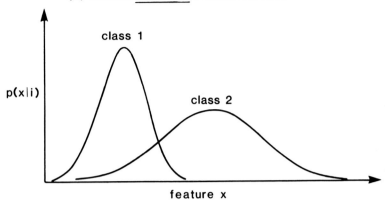

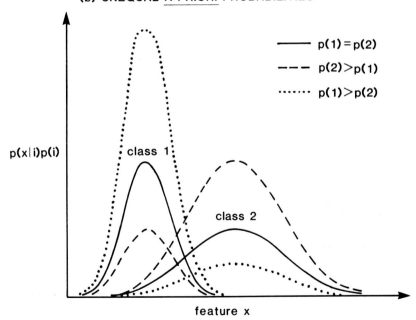

FIGURE 1-16. *The effect of the <u>a priori</u> probability on class probability <u>density</u> functions.*

functions, $p(x|i)p(i)$, represent the probability that a pixel has a feature value x *and* is in class i. In remote sensing the *a priori* probabilities may be estimated from external sources of information about the scene such as ground surveys, existing maps, or historical data.

To make a classification decision for a pixel, we need to know the *a posteriori* probabilities that the pixel belongs to each of the training classes, given that the pixel has the feature value x. This probability, $p(i|x)$, may be calculated with *Bayes Rule*

$$p(i|x) = p(x|i)p(i)/p(x) \qquad (1\text{-}23)$$

where
$$p(x) = \sum_{i=1}^{2} p(x|i)p(i) \qquad (1\text{-}24)$$

A *decision rule* may now be formed with the *a posteriori* probabilities of Eq. (1-23). If a pixel has feature value x, an intuitively satisfying approach is to assign the pixel to class 1 if $p(1|x)$ is greater than $p(2|x)$. Similarly, the pixel would be assigned to class 2 if $p(2|x)$ is greater than $p(1|x)$. Since $p(x)$ is the same for both classes in Eq. (1-23) it can be ignored in a comparison of the two, and we can write as the *Bayes decision rule*

a pixel belongs to class 1 if $p(x|1)p(1) > p(x|2)p(2)$

a pixel belongs to class 2 if $p(x|2)p(2) > p(x|1)p(1)$

In the very unlikely situation that the two *a posteriori* probabilities are exactly equal, i.e.

$$p(1|x) = p(2|x)$$

or

$$p(x|1)p(1) = p(x|2)p(2) \tag{1-25}$$

a decision cannot be made from the class probabilities. A tie-breaking process then must be employed, such as using the classification of an adjoining, previously classified pixel or randomly choosing either class 1 or class 2. It can be shown (Duda and Hart, 1973) that the Bayes decision rule minimizes the average probability of error over the entire classified data set, if all the classes have *normal* (Gaussian) probability density functions.

In practice, reliable *a priori* probabilities are difficult to obtain and, consequently, they are commonly assumed to be equal (to 0.5 in the two-class case). More accurate classification should result, however, if they can be accurately estimated from external data. If, for example, the goal is to determine the proportion of crop types planted during a particular season from Landsat images of an agricultural area, we might reasonably set the *a priori* probabilities equal to historical estimates of the percentage of each crop type in the area. A discussion of the use of *a priori* probabilities in remote sensing is given by Strahler (1980).

1.5.2 Discriminant Functions

The Bayes decision rule may be restated as

a pixel belongs to class 1 if $D_1(x) > D_2(x)$

a pixel belongs to class 2 if $D_2(x) > D_1(x)$

where $D_i(x)$ is called a *discriminant function* and is given by

$$D_i(x) = p(x|i)p(i) \qquad (1\text{-}26)$$

Note the crossover point, x_D, of the two functions in Fig. 1-17. This point is a *decision boundary* , or class partition; to the right of the boundary the decision favors class 2 and to the left of the boundary the decision favors class 1. Setting D_i equal to the *a posteriori* probabilities, Eq. (1-26), results in a Bayes optimal classification, but is not the only choice that has the same result.

Other discriminant functions may be derived by noting that the decision boundary is the same if any *monotonic* function of D is used. For example

$$D_i(x) = a[p(x|i)p(i)] + b \qquad (1\text{-}27)$$

or $\qquad\qquad D_i(x) = \ln[p(x|i)p(i)] \qquad (1\text{-}28)$

are both valid discriminant functions. The latter transformation is particularly useful if the class probability distributions are normal, i.e.

$$p(x|i) = \frac{1}{\sqrt{2\pi\sigma_i^2}} \; \exp\left[-\frac{(x-\mu_i)^2}{2\sigma_i^2} \right] \qquad (1\text{-}29)$$

where $\qquad\qquad \mu_i$ = mean of x for class i

$\qquad\qquad\quad \sigma_i^2$ = variance of x for class i

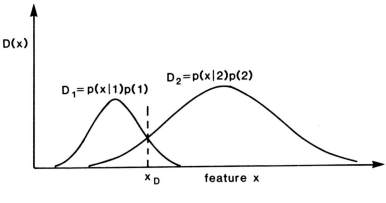

(a) THE SIMPLEST FORM

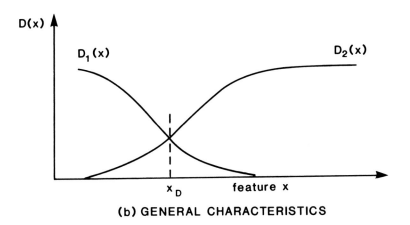

(b) GENERAL CHARACTERISTICS

FIGURE 1-17. Discriminant functions for the Bayes optimal partition between two classes.

A Bayes optimal discriminant function for class i is then

$$D_i(x) = \ln[p(x|i)p(i)]$$

$$= \ln[p(i)] - \frac{1}{2}\ln[2\pi] - \frac{1}{2}\ln[\sigma_i^2] - \frac{(x - \mu_i)^2}{2\sigma_i^2} \qquad (1-30)$$

which is a computationally efficient form because the last term, the only one that depends on x and, hence, the only one that must be recalculated at each pixel, is a simple quadratic function.

To see that this particular discriminant function is Bayes optimal, note that the two-class decision boundary is found by setting

$$D_1(x) = D_2(x) \qquad (1-31)$$

and solving for x. But this is equivalent to setting

$$\ln[p(x|1)p(1)] = \ln[p(x|2)p(2)] \qquad (1-32)$$

or $$\qquad p(x|1)p(1) = p(x|2)p(2) \qquad (1-33)$$

which corresponds to the crossover point of the *a posteriori* distributions (Fig. 1-17a), i.e., the discriminant function of Eq. (1-28) yields a Bayes optimal classification.

The total probability of classification error is given by the area under the overlapping portions of the *a posteriori* probability functions as shown in Fig. 1-18. The total probability of error is the sum of the probabilities that an incorrect decision was made on either side of the class partition. It is easy to see that the Bayes optimal partition minimizes this error because a shift of the partition to the right or left will include a larger area from either class 2 or class 1, respectively, thus increasing the total error.

It is instructive at this point to note again the role of the *a priori* probabilities. From Fig. 1-16b we see that the decision boundary will move to the left if p(2) is greater than p(1) and to the right if p(1) is greater than p(2). Even if reasonable estimates of the *a priori* probabilities are available, we may choose to bias them heavily if the significance of an error for one class is much greater than for the others. For example, the entire purpose of a hypothetical project may be to locate all occurrences of a rare class. The actual *a priori* probability of that class would be very low, but we could assign an artificially high *a priori* probability to insure that no occurrences are missed. The increased number of false identifications for the rare class then would have to be removed by site visits or by referencing other data, such as aerial photography. Obviously, if the analyst chooses to use non-equal *a priori* probabilities, they must be applied with considerable care and an appreciation of their importance in the classification process.

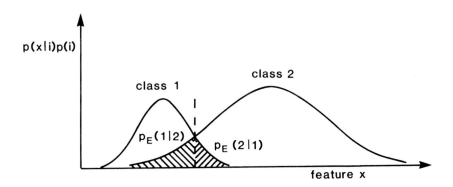

FIGURE 1-18.   Probability of error, p_E, for a maximum-
likelihood classification.

1.5.3 Extension to K Dimensions

The extension of the above concepts to K dimensions (and more than two classes) is straightforward, although the resulting mathematics is more complex. The data measurement variable, x, which to this point has represented the one-dimensional feature value of a pixel in a black and white image, becomes a measurement *vector*, X, having K components (Fig. 1-19) that represents, for example, a set of gray levels for a pixel in a multispectral image set. The probability functions, $p(x|i)$, become *multivariate* functions, $p(X|i)$, and the class partitions become curved lines in two dimensions, surfaces in three dimensions, and hypersurfaces in K dimensions.

As an example of this dimensionality extension, consider its effect on the normal distribution. In one dimension the normal distribution is given by Eq. (1-29) and requires only two parameters, the class mean, μ, and variance, σ^2, to specify the function completely. Similarly, the only parameters of a two-dimensional normal distribution are the class mean vector, M, and covariance matrix, Σ, the two-dimensional analog of the one-dimensional variance. Figure 1-20 illustrates the parameters of a two-dimensional normal distribution. The projection of the distribution onto each of the two feature axes yields the two class means, μ_1 and μ_2, which constitute the components of the class mean vector, and the class variances, σ_{11} and σ_{22}, which constitute the diagonal terms of the covariance matrix. The remaining elements of the covariance matrix, σ_{12} and σ_{21}, are calculated from a sample of N pixels in class i by

$$\sigma_{12i} = \sum_{\ell=1}^{N} [x_1(\ell) - \mu_{1i}][x_2(\ell) - \mu_{2i}]/(N - 1) \qquad (1-34)$$

where $x_1(\ell)$ and $x_2(\ell)$ are the two feature values of sample ℓ.

(a) ONE PIXEL SAMPLE

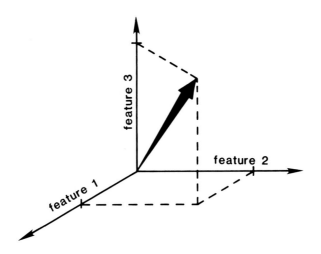

(b) MULTIPLE PIXEL SAMPLES

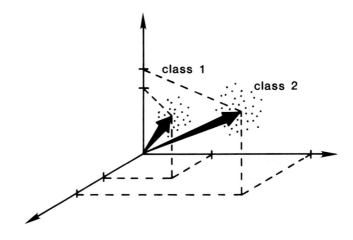

FIGURE 1-19. *Pixel vectors in three dimensions.*

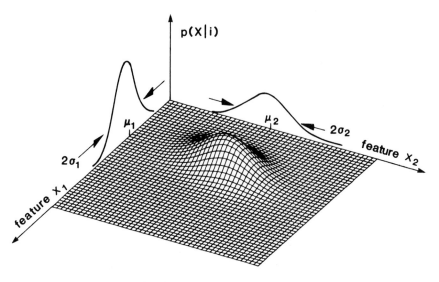

mean vector $M = \begin{pmatrix} \mu_1 \\ \mu_2 \end{pmatrix}$

covariance matrix $\Sigma = \begin{pmatrix} \sigma_1^2 & \sigma_{12} \\ \sigma_{21} & \sigma_2^2 \end{pmatrix}$

correlation coefficient $\rho = \sigma_{12} / (\sigma_1 \sigma_2)^{1/2}$

FIGURE 1-20. Two-dimensional normal distribution parameters.

The complete covariance matrix in two dimensions is given by

$$\Sigma_i = \begin{pmatrix} \sigma_{11i} & \sigma_{12i} \\ \sigma_{21i} & \sigma_{22i} \end{pmatrix}$$

$$= \begin{pmatrix} \sigma_{1i}^2 & \sigma_{12i} \\ \sigma_{21i} & \sigma_{2i}^2 \end{pmatrix} \tag{1-35}$$

Note that σ_{12i} equals σ_{21i}, i.e., the covariance matrix is symmetric, a property that is also true for higher dimensions. Also note that because the diagonal elements are the variances of the distribution along each dimension, e.g., $\sigma_{11i} = \sigma_{1i}^2$, they are always positive; however, there is *no* positive constraint on the off-diagonal elements.

The significance of the off-diagonal terms of the covariance matrix may be appreciated by defining the *correlation coefficient* between two dimensions as

$$\rho_{12i} = \sigma_{12i}/(\sigma_{11i}\sigma_{22i})^{1/2} \qquad (1-36)$$

In this normalized form, ρ_{12i} must have a value between minus one and plus one. Examples of the shape of a two-dimensional normal distribution for different values of ρ_{12i} are shown in Fig. 1-21. Note that values of ρ_{12i} close to plus or minus one imply a strong linear dependence between the data in the two dimensions, whereas if ρ_{12i} is near zero there is little dependence between the two dimensions. We will see later (Sec. 3.4.3) how the off-diagonal elements in the covariance matrix may be changed to zero by appropriate transformation of the K-dimensional image. The K features of the transformed image are therefore uncorrelated, a useful property for data analysis.

The general multivariate form for an K-dimensional normal distribution is

$$p(X|i) = \frac{1}{|\Sigma_i|^{1/2}(2\pi)^{K/2}} \exp[-1/2(X - M_i)^t \Sigma_i^{-1}(X - M_i)] \qquad (1-37)$$

where   X = pixel feature vector

M_i = mean vector for class i

Σ_i = K x K symmetric covariance matrix for class i

$|A|$ = determinant of matrix **A**

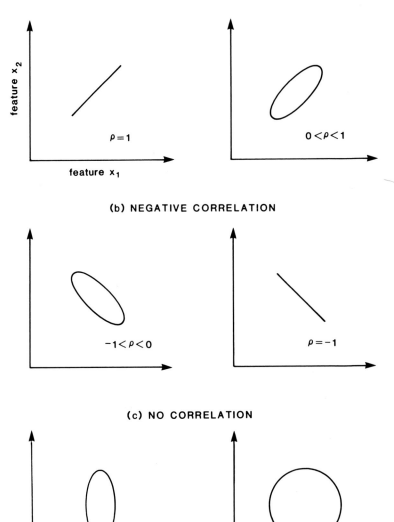

FIGURE 1-21. The effect of the correlation coefficient on the shape of a two-dimensional normal distribution.

A^t = transpose of matrix A

and A^{-1} = inverse of matrix A

In K dimensions the Bayes optimal discriminant functions become

$$D_i(X) = \ln[p(i)] - \frac{K}{2}\ln(2\pi) - \frac{1}{2}\ln|\Sigma_i| - \frac{1}{2}(X - M_i)^t \; \Sigma_i^{-1}(X - M_i)$$

$$(1\text{-}38)$$

as given by Duda and Hart (1973). The four terms in Eq. (1-38) are analogous to those in Eq. (1-30). The maximum-likelihood class partitions defined by Eq. (1-38) are hyperquadrics whose specific shape depends on the relations between the means and covariance matrices of the different classes. Duda and Hart (1973) plot examples of these partitions for discrimination between two bivariate normal distributions of different shapes and relative positions.

1.5.4 The Minimum-Distance Classifier

A commonly used algorithm for image classification is the minimum-distance classifier. With this algorithm, each unknown pixel with feature vector X is classified by assigning it to the class whose mean vector, M_i, is closest to X. In addition to the obvious intuitive appeal and computational simplicity of this approach, it can be shown that it is a very special case of the general maximum-likelihood classifier. If we assume that the covariance matrices of all the candidate classes are equal, i.e.

$$\Sigma_i = \Sigma_j = \Sigma_0 \qquad (1\text{-}39)$$

and that the *a priori* probabilities are equal

$$p(i) = p(j) = p_0 \qquad (1\text{-}40)$$

then the discriminant function of Eq. (1-38) becomes

$$D_i(X) = A - \frac{1}{2}(X - M_i)^t \Sigma_0^{-1}(X - M_i) \qquad (1\text{-}41)$$

where the constant A is given by

$$A = \ln[p_0] - \frac{K}{2} \ln[2\pi] - \frac{1}{2} \ln| \Sigma_0| \qquad (1\text{-}42)$$

and may be ignored in a comparison of D_i for different classes. The quantity

$$d_{Mi} = - \frac{1}{2}(X - M_i)^t \Sigma_0^{-1}(X - M_i) \qquad (1\text{-}43)$$

is called the *Mahalanobis Distance*. The expansion of this term in Eq. (1-41) results in a quadratic equation in X, whose quadratic term is independent of i (Duda and Hart, 1973) and may therefore be combined with the constant A. Thus Eq. (1-41) may be reduced to a form that is *linear* in X, meaning that the D_is are hyperplanes (linear functions in two dimensions), in contrast to the hyperquadric functions of Eq. (1-38) for arbitrary Σ_i.

If the covariance matrices are further constrained to be diagonal, i.e., the features are uncorrelated, and to have equal variance along each feature axis, i.e.,

$$\Sigma_i = \begin{pmatrix} \sigma_0^2 & & \text{zero} \\ & \ddots & \\ \text{zero} & & \sigma_0^2 \end{pmatrix} \qquad (1\text{-}44)$$

then

$$D_i(X) = A - \frac{(X - M_i)^t(X - M_i)}{2\sigma_0^2} \qquad (1\text{-}45)$$

The quantity $(X - M_i)^t(X - M_i)$ is a scalar consisting of a sum of squared terms

$$d_{2i}^2 = (X - M_i)^t(X - M_i)$$
$$= (x_1 - \mu_{1i})^2 + (x_2 - \mu_{2i})^2 + \ldots + (x_K - \mu_{Ki})^2 \qquad (1\text{-}46)$$

It is therefore simply the square of the *Euclidian distance* between the vectors X and M_i. Equation (1-45), therefore, represents the discriminant functions for the miniumum–distance classifier, for $D_i(X)$ will be largest for that class i for which the distance d_{2i} is a minimum, i.e., the class with the nearest mean.

Another measure of distance, the "city block" distance, is sometimes used to provide a more efficient implementation of Eq. (1-45). The "city block" distance is given by

$$d_{1i} = |x_1 - \mu_{1i}| + |x_2 - \mu_{2i}| + \ldots + |x_K - \mu_{Ki}| \qquad (1\text{-}47)$$

These two distance measures are illustrated in Fig. 1-22 for K equal to 2.

To demonstrate the different decision boundaries created by the maximum-likelihood rule and the minimum–distance rule, with the two distance measures d_1 and d_2, a simulated set of three normal distributions in two dimensions is used in Fig. 1-23. The probability density functions have different means and different, non-diagonal covariance matrices, but the *a priori* probabilities are equal. The decision boundaries resulting from the different algorithms are shown in Fig. 1-23b. The following points are evident from this example:

(1) The maximum-likelihood and minimum–distance decision boundaries are entirely different. They would be identical *only* if the conditions leading up to Eq. (1-45) are satisfied.

(2) The minimum-distance algorithm yields different decision boundaries with the two distance functions, d_1 and d_2. The boundaries for d_1 are piecewise linear and approximate the linear boundaries for d_2.

(3) The detached region (upper left) assigned to class 2 in the maximum-likelihood case indicates that the probability for class 2 is greatest in that region. It also indicates the need for *threshold boundaries* (Sec. 3.6.1) to avoid unreasonable classification of outliers, i.e., pixels whose feature vectors are far from any of the defined class mean vectors.

In summary, the mimimum-distance rule results in a simple algorithm that can be programmed efficiently, particularly with

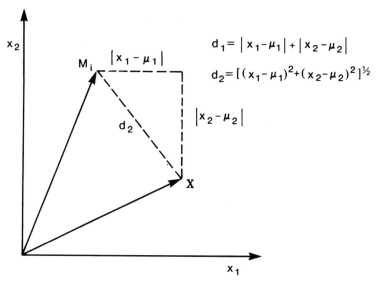

$$d_1 = \left| x_1 - \mu_1 \right| + \left| x_2 - \mu_2 \right|$$

$$d_2 = \left[(x_1 - \mu_1)^2 + (x_2 - \mu_2)^2 \right]^{\frac{1}{2}}$$

FIGURE 1-22.   The city block distance (d_1) and Euclidean distance (d_2) in two dimensions.

(a) SIMULATED NORMAL CLASS DISTRIBUTIONS

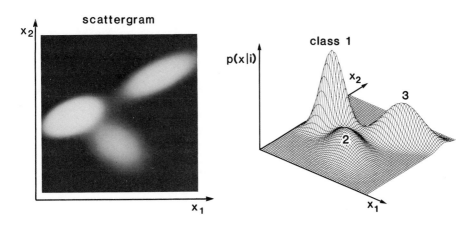

(b) CLASSIFICATION DECISION BOUNDARIES

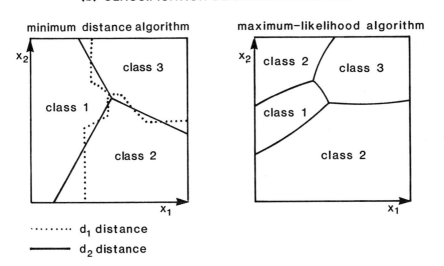

FIGURE 1-23. *Decision boundaries for a three-class, two-dimensional set of normal distributions.*

the d_1 distance measure. However, unlike the maximum-likelihood rule, it does *not*, in theory, minimize the average classification error, except in the very special case of diagonal class covariance matrices (the features are uncorrelated) with equal variances in all the features, *and* only if the distance measure d_2 is used. Nevertheless, one comparative study of classification algorithms in an agricultural application (Hixson et al, 1980) concluded that the minimum-distance algorithm (with distance measure d_2) yielded accuracies comparable to those from the maximum-likelihood algorithm in about one-half the computation time.

References

Anuta, Paul. E., "Spatial Registration of Multispectral and Multitemporal Digital Imagery using Fast Fourier Transform Techniques," IEEE Transactions on Geoscience Electronics, GE-8, No. 4, October 1970, pp. 353-368.

Biberman, L.M., "A Summary," Chap. 8 in Perception of Displayed Information, Biberman, ed., New York, Plenum Press, 1973, 345 pp.

Bracewell, R., The Fourier Transform and Its Application, 2nd ed., New York, McGraw-Hill, 1978, 444 pp.

Brigham, E. Oran, The Fast Fourier Transform, Englewood Cliffs, New Jersey, Prentice-Hall, 1974, 252 pp.

Churchill, Ruel V. and James W. Brown, Complex Variables and Applications, 3rd ed., New York, McGraw-Hill, 1974, 352 pp.

Churchill, Ruel V. and James W. Brown, Fourier Series and Boundary Value Problems, 3rd ed., New York, McGraw-Hill, 1978, 248 pp.

Cornsweet, T.N., Visual Perception, New York, Academic Press, 1970, 475 pp.

Duda, R.D. and P.E. Hart, Pattern Classification and Scene Analysis, New York, John Wiley and Sons, 1973, 482 pp.

Gaskill, Jack D., Linear Systems, Fourier Transforms and Optics, New York, John Wiley and Sons, 1978, 554 pp.

Goodman, Joseph W., Introduction to Fourier Optics, New York, McGraw-Hill, 1968, 287 pp.

Hixson, Marilyn, Donna Scholz, Nancy Fuhs, and Tsuyoshi Akiyama, "Evaluation of Several Schemes for Classification of Remotely Sensed Data," Photogrammetric Engineering and Remote Sensing, Vol. XLVI, No. 12, December 1980, pp. 1547-1553.

Legault, Richard, "The Aliasing Problems in Two-Dimensional Sampled Imagery," Chap. 7 in Perception of Displayed

Information, Biberman, ed., New York, Plenum Press, 1973, 345 pp.

Pratt, William K., Digital Image Processing, New York, John Wiley and Sons, 1978, 750 pp.

Slater, Philip N., Remote Sensing - Optics and Optical Systems, Reading, Mass., Addison-Wesley, 1980, 575 pp.

Strahler, Alan H., "The Use of Prior Probabilities in Maximum Likelihood Classification of Remotely Sensed Data," Remote Sensing of Environment, Vol. 10, 1980, pp. 135-163.

Swain, P.H. and S.M. Davis, eds., Remote Sensing: The Quantitative Approach, New York, McGraw-Hill, 1978, 396 pp.

CHAPTER 2

Digital Image Processing

2.1 Introduction

Digital processing of remote sensor imagery has become increasingly important since the advent of the Landsat satellite system. During the five and one-half years of operation of Landsat-1 (July 23, 1972, to January 6, 1978) approximately 272,000 multispectral scanner system (MSS) images, each containing four spectral bands, were acquired and cataloged by NASA and foreign stations. Each image is originally in digital form and contains a total of about 28 million pixels in the four spectral bands. The total amount of data generated during the satellite's lifetime was thus 7.6 *trillion* pixels, an average data rate of almost 44,000 pixels per second over the entire five and one-half years. NASA is currently developing computers that will process about 10^{10} pixels per day (116,000 pixels per second) to handle the large amount of data from the next generation of satellite sensors (Schafer and Fischer, 1982).

Although digital image processing cannot match the speed of optical image processing (Goodman, 1968), it is the logical choice for processing the large quantity of remote sensing imagery for several reasons:

(1) Digital processing of the original digital data has the greatest potential for preserving the correct radiometry and the maximum resolution of the images.

(2) Digital processing is more flexible than optical processing. This flexibility is limited, however, by such factors as the unavoidable compromise between

computation time (hence cost) and the accuracy or
sophistication of the results.

(3) The general availability of computers permits different
    investigators to perform repeatable and quantifiable
    analyses on common imagery.

Because of the large amount of image data involved in
remote sensing, the application of digital image processing has
developed in two different contexts. The correction of systematic errors and calibration of the imagery is called *preprocessing* and requires that relatively simple, repetitive operations
be performed on large numbers of images to create a consistent
and reliable imagery data base. Consequently, the primary
computational requirements are high speed software algorithms
and input/output (I/O) devices. On the other hand, *information
extraction* from the preprocessed images requires maximum flexibility and a user-oriented approach in software and display
devices, in addition to fast processing algorithms. The current
trend is to perform the preprocessing operations at large,
special purpose facilities such as the EROS Data Center and
Goddard Space Flight Center. The preprocessed data are then
disseminated to researchers for specialized processing.

In this chapter, techniques for preprocessing and enhancement of images are described, while the equally important areas
of automatic scene classification and thematic mapping are
discussed in Chapter 3, in accord with the distinction made in
Sec. 1.3. The emphasis in both chapters is on the processing
techniques, not specific remote sensing applications.

2.2 Contrast Manipulation

Contrast manipulation is a pixel-by-pixel radiometric
transformation that is designed to enhance visual discrimination
of low contrast image features. Each pixel's gray level is

changed by the specified transformation, without regard for neighboring pixel gray levels (Fig. 1-8). The pixel-by-pixel transformation property allows contrast modification to be performed by a table look-up procedure, a fast processing technique well-suited to the integer arithmetic often used on minicomputers and to hardware implementation in interactive CRT display systems (Appendix C).

The examples of contrast modification in this section illustrate why it is often beneficial to alter the true radiometric relationships between pixels in an image that will be analyzed visually. Generally nothing is gained, however, by applying the same processing techniques to images destined for classification by computer. In fact, any alteration of the original radiometric values in an image that will later be classified should be avoided unless the transformations are known to be beneficial (Sec. 3.4).

2.2.1 Gray Level Histograms

Examination of the image *histogram* is a useful, and often necessary, preliminary step for successful manipulation of image contrast. The image histogram describes the statistical distribution of gray levels in an image in terms of the number of pixels (or percentage of the total number of pixels) having each gray level. It is analogous to the familiar probability density function in statistics. Histograms for many images tend to be Gaussian in shape, but often have an extended tail toward higher gray levels (higher scene radiances). The general characteristics of gray level histograms for a variety of images are illustrated in Fig. 2-1. It is important to remember that an image histogram only specifies the total number of pixels at each gray level; it contains no information about the *spatial* distribution of gray levels throughout the image.

The effect of a gray level transformation on the image histogram is shown in Fig. 2-2. Each gray level in the input image is projected through the transformation curve to a new gray level in the output image. Note that the *area* of the histogram is not changed by the gray level transformation; the area is always the total number of pixels in the image (or 100 percent if the histogram is normalized). The simple linear transformation of Fig. 2-2 is commonly called a *contrast stretch* and is accomplished by first subtracting a bias gray level, and

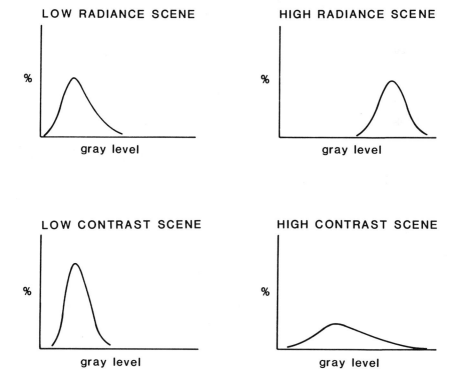

FIGURE 2-1. *Histogram characteristics for different types of scenes.*

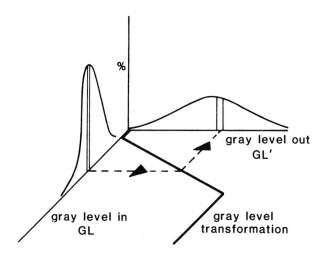

FIGURE 2-2. Gray level transformation.

then increasing the gray level range with a gain factor. This transformation is ideally implemented on an interactive CRT image display system with a two-parameter feedback, such as the x-y position of an operator-controlled CRT cursor (Appendix C).

2.2.2 Contrast Enhancement

Examples of *linear* radiometric transformations and the effect each has on the image histogram are depicted in Fig. 2-3. The simple linear transformation is routinely used to increase the contrast of a displayed image by expanding the original gray level range to fill the dynamic range of the display device. To achieve a greater contrast increase, we can usually accept some saturation at both extremes of the output range, unless important image structure is lost in the saturated

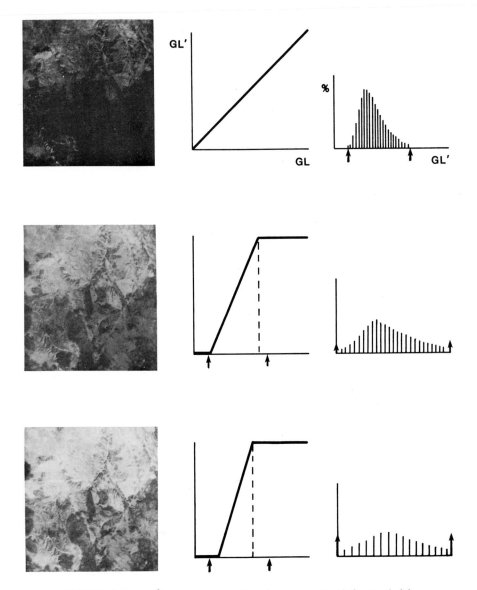

FIGURE 2-3. Linear contrast enhancement with variable saturation.

areas of the image. This type of transformation also can be used to *decrease* image contrast when the image gray level range exceeds that of the display.

If the image histogram is asymmetric, as it often is, it is impossible to simultaneously control the average gray level of the output image and the amount of saturation at the ends of the histogram with a simple linear transformation. With a two-segment *piecewise linear* transformation, more control is gained over the image contrast and the histogram asymmetry can be reduced, thus making better use of the available gray level range (Fig. 2-4a). More than two linear segments may be used in the transformation, of course, with a corresponding increase in control of the processed image's contrast.

A more general nonlinear transformation, called *histogram equalization*, is useful for a wide variety of images. If the *cumulative distribution function* of the original image (the total number of pixels in the histogram between zero and each gray level) is used as the transformation function, after appropriate scaling of the ordinate axis to correspond to output gray levels, the histogram of the processed image will be approximately uniform (Gonzalez and Wintz, 1977). Histogram equalization tends to *automatically* reduce the contrast in very light or dark areas and to expand the middle gray levels toward the low and high ends of the radiance scale because of the Gaussian shape of most image histograms. The contrast of the processed image is often rather harsh, a characteristic that may be offset by the fact that no parameters are required from the analyst to implement the transformation. An example of histogram equalization is shown in Fig. 2-4b.

A *cyclic* transformation divides the full input range of gray levels into several smaller portions, each of which is subjected to a similar transformation. An example of this type of processing is shown in Fig. 2-5. The very low contrast

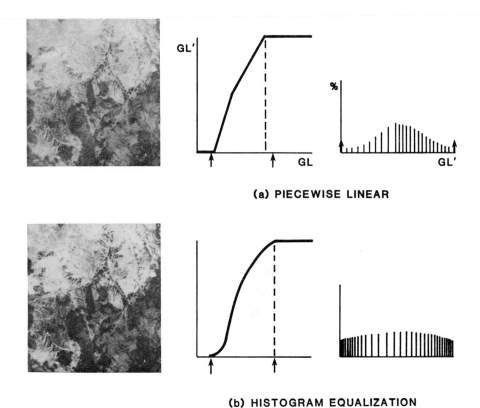

(a) PIECEWISE LINEAR

(b) HISTOGRAM EQUALIZATION

FIGURE 2-4. Nonlinear contrast enhancement.

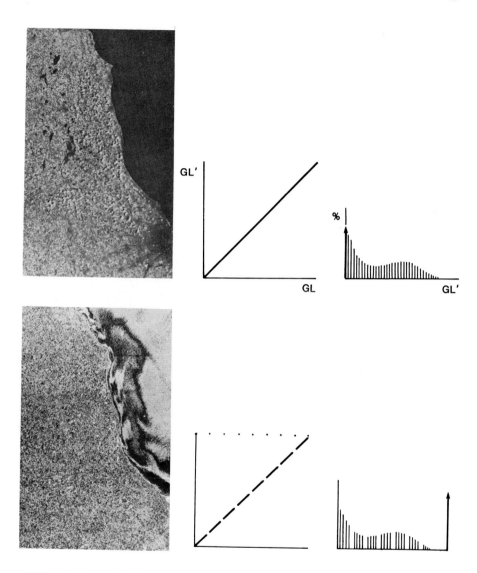

FIGURE 2-5. Cyclic contrast enhancement (Blackwell and Boland, 1975; reproduced with permission from Proc. Am. Soc. of Photogrammetry Annual Convention, March 1975, ©1975, Am. Soc. of Photogrammetry).

structure in the coastal waters has been contoured by the periodic spikes in the transformation, thereby rendering it easily visible. In this case, the land area has also been processed in the same way, but, because the structure there is fine and of relatively high contrast, it has been transformed into uninterpretable noise.

The degradation of the land areas in Fig. 2-5 is an example of the major drawback of the cyclic transformation, i.e., it does not transform each original gray level to a unique output gray level, but transforms several different gray levels into a single gray level (Fig. 2-5). Once the transformation has been performed there is no way to extract the original gray levels, i.e., the transformation cannot be reversed. A solution to this problem for images such as the one in Fig. 2-5 is to apply a different transformation to the land and water areas. The boundary between land and water, therefore, must be accurately delineated to create masks for isolating the effect of each transformation. The near infrared bands of multispectral sensors are useful for this purpose because water has nearly zero reflectance at those wavelengths, and is easily separated from land or vegetation by gray level thresholding (Sec. 2.2.3).

All of the contrast enhancement examples discussed thus far have been *global* implementations, i.e., the same contrast stretch was applied to all image pixels. Obviously, contrast can vary locally within an image; a more optimal enhancement may therefore be achieved by using an *adaptive* algorithm whose parameters change from pixel-to-pixel according to local image contrast (Fahnestock and Schowengerdt, 1983). Figure 2-6 is an example of adaptive contrast enhancement. In this case the image was partitioned into 64 adjoining blocks, or subimages. The minimum and maximum gray levels in each subimage were found and used to specify a different linear stretch function for each subimage. The parameters for the transformation at each pixel

were then calculated, pixel-by-pixel, by linear interpolation of
the stretch parameters in adjoining subimages to provide a
spatially smooth contrast enhancement, without sharp discontinuities at the boundaries between subimages. Although the global
radiometry of the image is changed by the processing, it is much
easier to distinguish previously low contrast features in the
darker and brighter portions of the image.

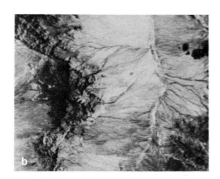
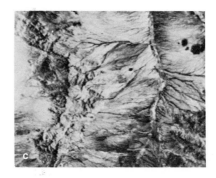

FIGURE 2-6.   Local contrast enhancement. (a) No stretch.
(b) Global stretch. (c) Local stretch.

2.2.3 Thresholding

Thresholding is a type of contrast manipulation that is *not* designed to enhance contrast, at least in the sense described above. Instead, it *segments* an image into two classes defined by a single gray level threshold. The use of a binary threshold on certain types of images results in sharply defined spatial boundaries that may be used for masking portions of the image. Separate processing may then be applied to each of the two classes and the results recombined to alleviate the difficulties encountered with images such as the one in Fig. 2-5. Thresholding may also be used as a simple classification algorithm, for example in a decision tree classifier (Sec. 3.5.2).

Figure 2-7 illustrates the application of gray level thresholding to segment land and water areas in a Landsat MSS band 7 image. Because the land area is densely vegetated, there is a wide separation between the water and land modes of the gray level histogram, making threshold selection relatively easy. In many cases, however, the two classes to be separated are much more similar and have a continuum of gray levels in the transition region. This more difficult situation may require the more sophisticated multispectral classification techniques discussed in Chapters 1 and 3.

Application of thresholding to the detection of change in a pair of multitemporal images is shown in Figs. 2-8 and 2-9. The area shown is a copper mining complex viewed in 1972 and 1975 with the Landsat MSS. The two images were first registered using control points and geometrical transformation (Sections 2.5 and 2.6). The registered images were then subtracted, pixel-by-pixel, to produce an image of *changes in gray level* from one date to the other. The distribution of these gray level changes is typically positive and negative about a gray

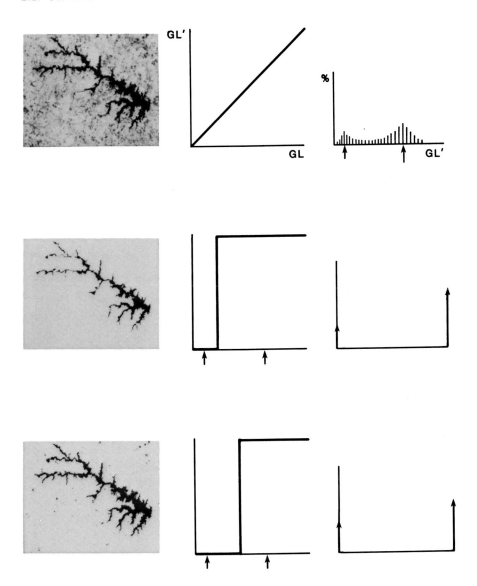

FIGURE 2-7. Binary thresholding for image segmentation.

FIGURE 2-8.   Two registered multitemporal Landsat images of
              a copper mining complex. (a) 1972. (b) 1975.

level of zero (if the two original images have equal mean gray
levels).   The difference image, therefore, must be adjusted to
all-positive gray levels for display as in Fig. 2-9a.

To *detect* (but not necessarily *identify*) changes between
the two images, the difference image may be thresholded to
clearly display gray level changes that are greater than a
certain magnitude (Figs. 2-9b and 2-9c).   Gray level thresholding can be easily implemented in an interactive mode with a
CRT display and operator-controlled cursor, but selection of the
"best" threshold level is, of course, a difficult task and must
usually be associated with *a priori* knowledge about the scene or
visual interpretation to be meaningful.   There has been
research in techniques for the quantitative determination of
optimum threshold levels (see Pun, 1981, for example) using only
information from the image histogram.

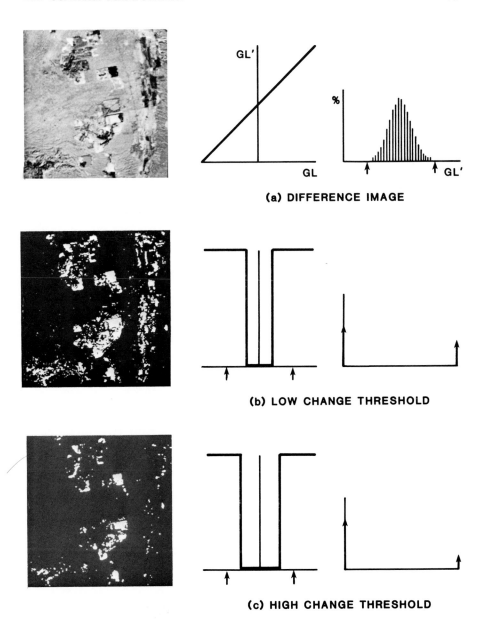

(a) DIFFERENCE IMAGE

(b) LOW CHANGE THRESHOLD

(c) HIGH CHANGE THRESHOLD

FIGURE 2-9.   Binary thresholding for change detection using the images of Fig. 2-8.

2.3 Spatial Filtering

Spatial filtering, like contrast manipulation, is a pixel-by-pixel transformation of an image. The transformation in this case, however, depends not only on the gray level of the pixel being processed, but also on the gray levels of neighboring pixels. Therefore, spatial filtering is a *context-dependent* operation that alters the gray level of a pixel according to its relationship with the gray levels of other pixels in the immediate vicinity. This conceptual difference between contrast manipulation and spatial filtering is shown in Fig. 1-8. Because spatial filtering uses several pixels to transform each pixel, it generally requires more calculations than contrast enhancement. For certain simple types of PSFs, a very efficient algorithm may be used to reduce the computational burden. In addition, recent advances in image processing hardware architecture (Reader and Hubble, 1981) have made spatial filtering possible in a nearly interactive environment.

2.3.1 Types of Filters

Spatial filters used in image processing are based on three basic types, *low*, *high*, or *band-pass* (Fig. 2-10). These basic filters may be additively combined to form a wide variety of more complex filters. Pure band-pass filters do not have general application in image processing; their primary use is for isolating periodic noise from an image (Sec. 2.4.3). In image processing, spatial filters are two-dimensional functions, as shown in Fig. 1-13. If the MTF (Sec. 1.4.3) of a filter is greater than one for certain values of spatial frequency (ν_x, ν_y), the filter *boosts*, i.e., increases, the amplitude of those spatial frequency components. The filter *attenuates* those frequency components where the MTF is less than one.

The effects of low and high-pass filtering on an image and its histogram are shown in Fig. 2-11. Low-pass filtering smooths the detail in an image and reduces the gray level range (i.e., the image contrast) and high-pass filtering enhances detail at the expense of large area radiometry and produces an image with a relatively narrow histogram centered at zero gray level. A simple high-pass filter may be implemented by sub-tracting a low-pass filtered image from the original, unproces-sed image, or by convolution with a PSF having a positive weight for the center pixel and negative weights for surrounding pixels. The calculation within the moving PSF window, therefore, involves the difference between the central pixel and

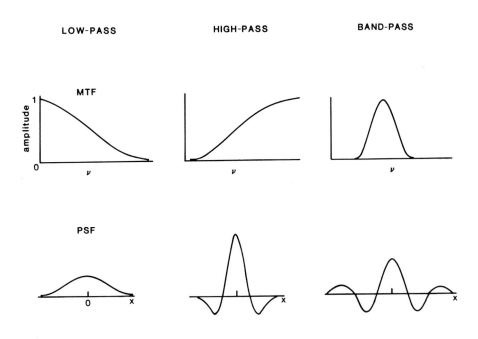

FIGURE 2-10. The three basic types of spatial filters.

histogram

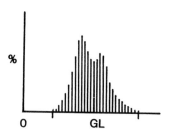

(a) ORIGINAL IMAGE

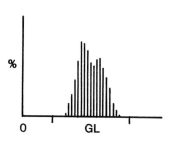

(b) LOW-PASS IMAGE

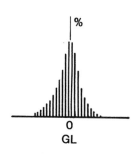

(c) HIGH-PASS IMAGE

FIGURE 2-11.   *Characteristics of spatially filtered images.*
(Schowengerdt, 1980; reproduced with permission
from Photogrammetric Eng. and Remote Sensing,
October, 1980, ©1980, Am. So. of Photogrammetry)

surrounding pixels, and because negative differences are just as likely to occur as positive differences, the histogram of high-pass images is virtually always symmetric about a mean gray level of zero. High-pass images, therefore, must be contrast stretched to all-positive gray levels for display.

Examples of low- and high-pass filtering of a Landsat image with different window sizes are shown in Fig. 2-12. All the processed images have been linearly contrast stretched to emphasize their differences. As the window size increases, the low-pass images become more blurred (exactly analogous to the effect of defocus in an optical system) because more higher frequency components are removed or attenuated in the image. On the other hand, as the window size increases in the high-pass images, fewer low frequency components are subtracted from the original, while more higher frequency components are retained. Very large windows (101-by-101 pixels, for example) may be used to selectively remove the very lowest frequency components from an image. Such processing is one way to remove the large area shading often present in vidicon images.

A *high-boost* filter may be created by using a weight K on the original image in a high-pass filter calculation

$$
\begin{aligned}
\text{high boost} &= (K)\text{original} - \text{low pass} \\
&= (K-1)\text{original} + \text{original} - \text{low pass} \\
&= (K-1)\text{original} + \text{high pass} \qquad (2\text{-}1)
\end{aligned}
$$

The result, therefore, is to partially restore the low frequency components that are lost in a high-pass operation. A standard high-pass image results if K equals one. For values of K greater than one, the processed image looks more like the original image, with a degree of edge enhancement (Fig. 2-13) that is dependent on K (Wallis, 1976; Lee, 1980). The parametric nature of this processing allows convenient "tuning" of K

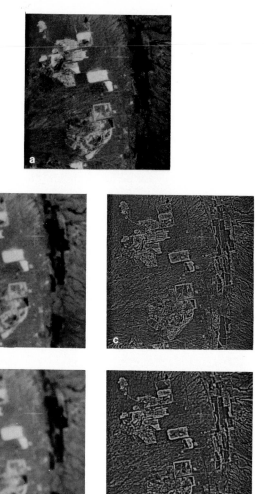

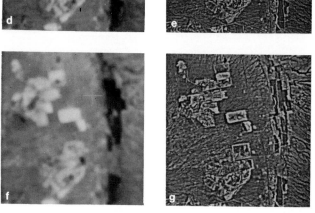

FIGURE 2-12.   Low- and high-pass filtering with different size
PSFs. Left page: (a) original, (b-g) processed
images (low pass in left column, high pass in right

low-pass

$\frac{1}{9}$ ×

1	1	1
1	1	1
1	1	1

high- pass

$\frac{1}{9}$ ×

-1	-1	-1
-1	8	-1
-1	-1	-1

$\frac{1}{25}$ ×

1	1	1	1	1
1	1	1	1	1
1	1	1	1	1
1	1	1	1	1
1	1	1	1	1

$\frac{1}{25}$ ×

-1	-1	-1	-1	-1
-1	-1	-1	-1	-1
-1	-1	24	-1	-1
-1	-1	-1	-1	-1
-1	-1	-1	-1	-1

$\frac{1}{49}$ ×

1	1	1	1	1	1	1
1	1	1	1	1	1	1
1	1	1	1	1	1	1
1	1	1	1	1	1	1
1	1	1	1	1	1	1
1	1	1	1	1	1	1
1	1	1	1	1	1	1

$\frac{1}{49}$ ×

-1	-1	-1	-1	-1	-1	-1
-1	-1	-1	-1	-1	-1	-1
-1	-1	-1	-1	-1	-1	-1
-1	-1	-1	48	-1	-1	-1
-1	-1	-1	-1	-1	-1	-1
-1	-1	-1	-1	-1	-1	-1
-1	-1	-1	-1	-1	-1	-1

column); (b, c) 3 × 3, (d, e) 5 × 5, (f, g) 7 × 7.
Right page: PSF weights.

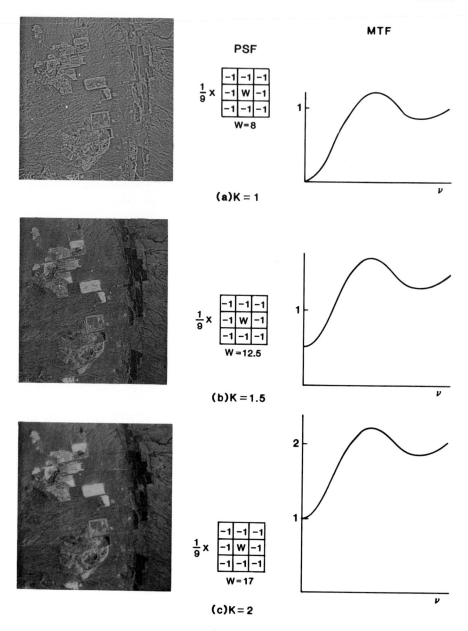

FIGURE 2-13. High-boost filtering. The center weight in
the PSF is calculated as W = 9*K - 1.

to achieve the degree of enhancement desired and would be ideally implemented in an interactive mode. This type of high frequency enhancement is currently available in computer-enhanced Landsat image products from the EROS Data Center (Holkenbrink, 1978).

A useful application of spatial filtering is *directional enhancement* of image features. Figure 2-14 depicts the results of preferential processing for vertically or diagonally-oriented features. These directional high-pass filters were designed to accentuate geologic lineaments and faults in the two directions. Note the complete disappearance of the road running to the northeast in the diagonal enhancement because it is orthogonal to the enhanced direction. Considerable care obviously must be exercised in interpreting spatially-filtered images, particularly directional enhancements, because of the abstract nature of the processed images.

2.3.2 The Box-Filter Algorithm

The simplest way to implement low and high-pass filters is by spatial neighborhood averaging. A low-pass filter can be implemented, for example, by averaging the pixels in the vicinity of each pixel of the original image, and using the average as the pixel gray level in the processed image. This technique is sometimes called a *moving average* (Fig. 1-15) and can be programmed in a very efficient recursive form.

Consider a PSF comprised of a 3-by-3 pixel array, as shown in Fig. 2-15. The amplitude of the PSF is one-ninth to maintain the average image gray level during the processing. Now visualize a convolution of this PSF with the image along a given line of the image. A straightforward calculation of the average at each pixel would require eight additions. However, if we save and update the average of the pixels in each of the three columns of the PSF, the eight additions need to be computed only

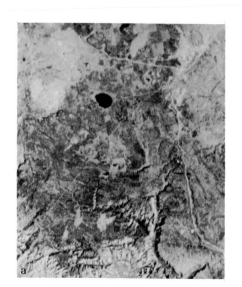

original

PSF | +1 | −1 |

| 0 | −1 |
| +1 | 0 |

vertical   diagonal

FIGURE 2-14.   Directional high-pass filtering (Schowengerdt
              et al, 1981; ©1981 Am. Water Resources Assoc.).

once at the beginning of an image line. Each subsequent average
can be calculated as the previous average (the output pixel to
the left in the same line) minus the old average in column 1,
plus the new average in column 3. Thus, except for the first
output pixel of each line, only three additions and one subtraction are needed to calculate each output pixel. The computational advantage of the recursive algorithm over the direct
calculation for a 3-by-3 PSF is therefore a factor of two. The
advantage obviously increases with the size of the PSF, and the

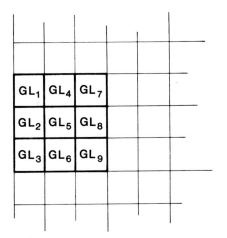

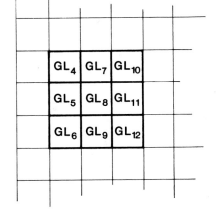

column sums:

$$C_1 = GL_1 + GL_2 + GL_3$$
$$C_2 = GL_4 + GL_5 + GL_6$$
$$C_3 = GL_7 + GL_8 + GL_9$$

output pixel:

$$GL'_5 = C_1 + C_2 + C_3$$

new column sum:

$$C_3 = GL_{10} + GL_{11} + GL_{12}$$

output pixel:

$$GL'_8 = GL'_5 - C_1 + C_3$$

(a) INITIALIZE ALGORITHM   **(b) RECURSION RELATION**

FIGURE 2-15. The box-filter algorithm.

recursion algorithm also can be applied in the vertical direction (McDonnell, 1981), further increasing its efficiency.

Many PSFs can be designed around the box-filter algorithm by changing the window size and combining weighted low and high-pass filters (Fig. 2-16). Maximum flexibility in spatial

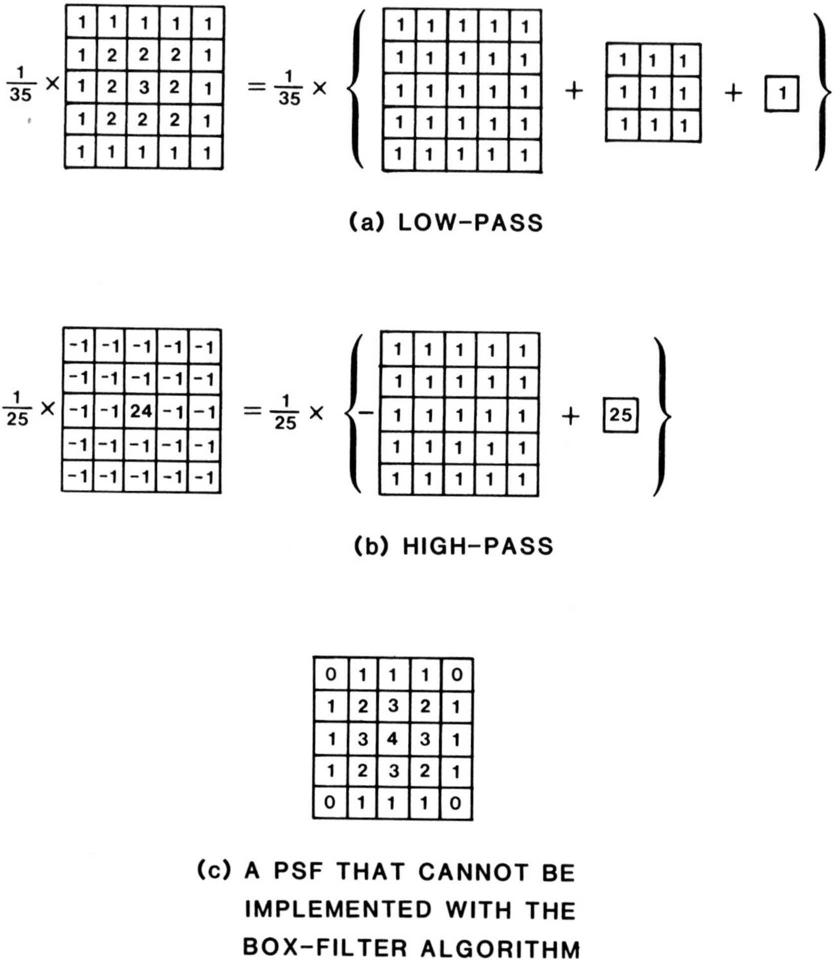

(a) LOW-PASS

(b) HIGH-PASS

(c) A PSF THAT CANNOT BE IMPLEMENTED WITH THE BOX-FILTER ALGORITHM

FIGURE 2-16. Application of the box-filter algorithm to more general PSFs.

filtering is achieved, however, by individually varying the weights of the PSF. The box-filter algorithm cannot be applied in this general situation. Fourier domain computation of linear filters then becomes a competitive alternative, particularly for large PSF windows. A general guide for choosing between spatial and Fourier domain processing with arbitrary filters is to use spatial domain convolution if the PSF is 7-by-7 or smaller and Fourier domain filtering otherwise (Pratt, 1978). Of course this choice will depend on the size of the image, the speed of the particular algorithms used, and the availability of special purpose hardware for performing convolutions or FFTs.

2.3.3 Edge Detection

A classic problem in image processing is the detection of sudden changes in gray level from one pixel to another. Such changes usually indicate a boundary, i.e., an edge, between two distinctly different objects in the image. Although many different approaches to this problem have been studied (Davis, 1975), a combination of high-pass spatial filtering and gray level thresholding provides a simple and effective technique that is widely used.

The directional high-pass filters of Fig. 2-14 produce images whose gray levels are proportional to the difference in neighboring pixel gray levels of the original image. However, the enhancement produced by a single filter is predominantly in one direction. Enhancement of edges in all directions may be achieved by filtering the image in two orthogonal directions, e.g., horizontally and vertically, and combining the results in a vector calculation (Fig. 2-17). The magnitude of the local image *gradient* is given by the length of the composite vector; the *direction* of the local gradient is given by the angle between the composite vector and the coordinate axis.

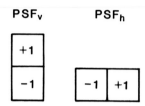

(a) VERTICAL AND HORIZONTAL FILTERS

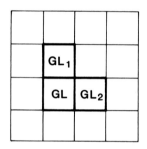

$$GL'_v = GL_1 - GL \qquad GL'_h = GL_2 - GL$$

(b) FILTER OUTPUTS

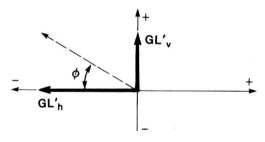

$$GL'_{mag} = \sqrt{(GL'_v)^2 + (GL'_h)^2}$$

$$\phi = \tan^{-1}(GL'_v / GL'_h)$$

(c) GRADIENT MAGNITUDE AND ORIENTATION

FIGURE 2-17. *Vector calculation of image gradients.*

Numerous gradient filters have been proposed (Fig. 2-18; Robinson, 1977). The output gradient image of the 2-by-2 filters is shifted by one-half pixel from the input image because the gradient is actually estimated between the original

0	−1
+1	0

−1	0
0	+1

(a) ROBERTS

+1	+2	+1
0	0	0
−1	−2	−1

−1	0	+1
−2	0	+2
−1	0	+1

(b) SOBEL

+1	+1	+1
+1	−2	+1
−1	−1	−1

−1	+1	+1
−1	−2	+1
−1	+1	+1

(c) PREWITT

FIGURE 2-18. PSF pairs for gradient filters.

pixels. The 3-by-3 filters do not produce this shift because of their symmetry about the central pixel. However, the edge boundaries produced by the larger filters are not as sharp as those produced by the 2-by-2 filters. Also note that, although the component horizontal and vertical filters are linear operations, the vector magnitude calculation is *not*. The overall gradient processing is therefore nonlinear. The performance characteristics of different edge detection filters, such as sensitivity to edge orientation and image SNR, have been studied by Pratt (1978).

The actual detection of edges in the gradient magnitude image is a binary classification problem that may be addressed with the gray level threshold technique described in Sec. 2.2.3. A gray level threshold applied to the gradient image results in *lines* at edge boundaries. A compromise threshold must usually be accepted because a threshold that is too low results in many isolated pixels identified as edges and thick, poorly defined edge boundaries, and a threshold that is too high results in thin, broken line segments. There are post-threshold processing techniques called *line thinning* and *connecting* that help alleviate these problems, but they are generally only partially successful. An example of edge detection using the Roberts filter on a Landsat MSS band 7 image is shown in Fig. 2-19. This image yields a relatively "clean" edge map because of the high contrast between the water and surrounding vegetation. Nevertheless, a few isolated pixels were accepted by the threshold as "edge" pixels.

2.3.4 OTF Correction

In Chapter 1 we described an image-forming optical system as a linear system. The performance of the optical system, in terms of image sharpness, can be described by the PSF in the spatial domain or the Optical Transfer Function (OTF) in the

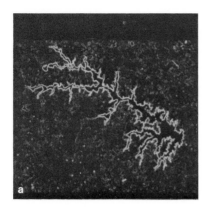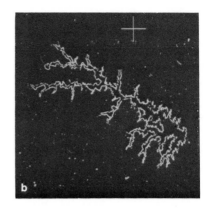

FIGURE 2-19.   Edge detection using the image of Fig. 2-7.
(a) Gradient. (b) Thresholded gradient.

spatial frequency domain.   The OTF is a low-pass filter that
attenuates the higher spatial frequency components in the scene,
thus blurring the image.   It is quite appealing to consider
undoing the effect of the OTF by dividing it into the image
spectrum, thereby achieving a "perfect" image (Fig. 2-20a).
This goal is unattainable in practice for two reasons: 1) those
frequency components for which the OTF is zero, including all
components above the cutoff frequency, are irrevocably lost in
the image; and 2) noise in the image from film grain or other
sources establishes an upper frequency limit, beyond which OTF
correction begins to degrade the resulting image.   This frequency limit is approximately at the point where the image modulation equals that of the noise, i.e., where the modulation SNR is
one.

In spite of these limitations, a considerable amount of
enhancement can be possible   with OTF correction as shown in

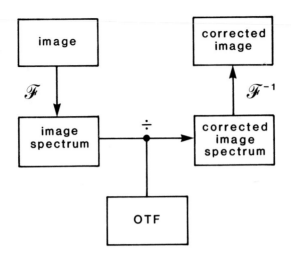

(a) PROCESSING STEPS

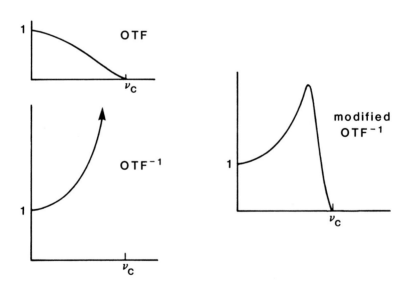

(b) HIGH FREQUENCY MODIFICATION

FIGURE 2-20. OTF correction.

Fig. 2-21. The primary source of degradation in this telescopic image is atmospheric turbulence. The higher resolution Lunar Orbiter image verifies that the processing has not introduced artifacts into the corrected image. Figure 2-22 shows the results of an OTF correction in the presence of two different noise levels. Although the image modulation is increased by the OTF correction in both cases, the noise is also amplified to the same degree. To alleviate this problem the inverse transfer function may be modified to attenuate the higher spatial frequencies, as indicated in Fig. 2-20b, thus suppressing the noise amplification (and, at the same time, some image detail).

The above description of OTF correction is a brief introduction to the much larger topic of *image restoration*. This subject has attracted many researchers, resulting in a large body of published literature describing many approaches to the problem that are more sophisticated than simple OTF correction. The book by Andrews and Hunt (1977) provides an excellent review and starting point for further reading on this topic.

2.4 Noise Suppression

Image noise is any unwanted signal or disturbance in an image. The most common example of image noise is photographic granularity. If the noise is severe enough to degrade the quality of the image, i.e., to impair extraction of information from the image, an attempt at noise suppression is warranted. Preliminary analysis of the image and noise structure is usually necessary before applying a noise suppression algorithm. In some cases, sensor calibration data and even test images may exist that are sufficiently comprehensive to estimate noise parameters. Unfortunately, however, these types of data are often incomplete or the noise is unexpected (for example, interference from other equipment or external signal sources). We

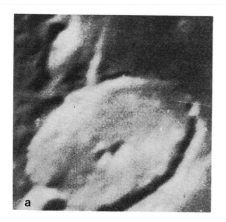 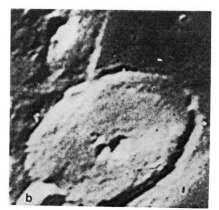

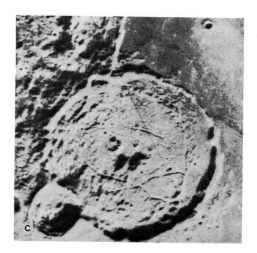

FIGURE 2-21. An example of OTF correction. (a) Telescopic image.
(b) Corrected image. (c) Lunar Orbiter image.
(O'Handley and Green, 1972; ©1972 IEEE)

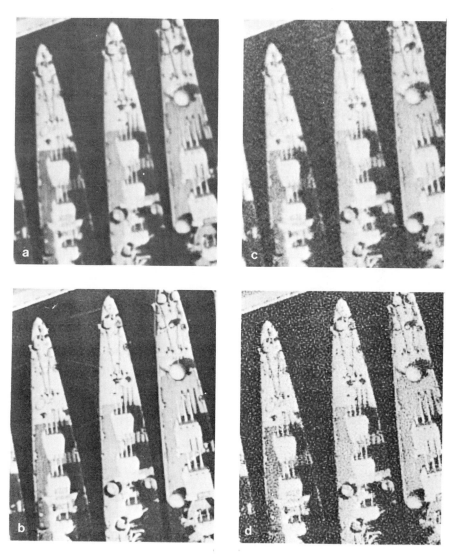

FIGURE 2-22.   OTF correction in the presence of noise. (a, b)
Original and corrected for high SNR. (c, d) Original
and corrected for low SNR. (Arguello, 1971)

then are faced with learning as much as we can about the noise from the noisy image itself. Periodic noise is relatively easy to diagnose and characterize from the imagery, but random noise is considerably more difficult to separate from valid image data. Trial-and-error processing of images with random noise is often the most expedient approach. The effectiveness of different processing algorithms on noise with unknown parameters is often evaluated by a simple judgment on the visual quality of the processed image. Image noise is divided into several distinct categories in the following discussion, but may occur as combinations of these types in actual images.

2.4.1 Random Noise

Random noise is characterized by a statistical variation in gray level from pixel to pixel. It may originate in the atmosphere, or electronic components, such as detectors or amplifiers, or it may be an inherent part of the image formation process, such as granularity in photographic images. Random noise that is independent of the irradiance (signal) level of the image incident on the detector is an assumption that is often made for simplifying mathematical models. Nevertheless, real noise commonly depends to some degree on the signal level.

Low-pass spatial filters can reduce random noise by averaging several pixels. If the noise is uncorrelated from pixel to pixel, its variance will be reduced by the low-pass filter. The obvious disadvantage of this technique is the image smoothing that occurs with the noise reduction. Many attempts have been made to develop more elaborate algorithms that simultaneously preserve image sharpness and suppress noise. An example is the so-called "noise cheating" algorithm devised by Zweig et al (1976).

A simple way to suppress random noise is to add several images of the same scene, each acquired independently. If the noise is random from image to image, it tends to average to zero in the addition process, thus increasing the image SNR without blurring the image. An example of noise suppression with this technique is shown in Fig. 2-23. Of course, the basic limitation to multiple image averaging is that we seldom have more

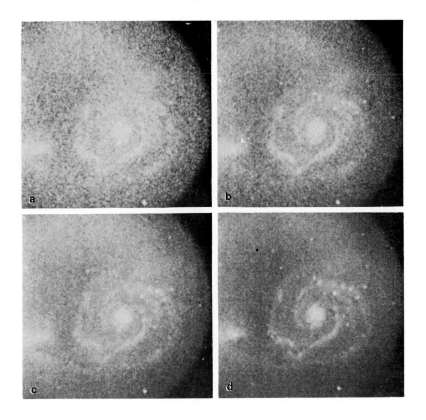

FIGURE 2-23.   *Random noise suppression by image averaging. (a)*
Single image. (b) Two averaged. (c) Four averaged.
(d) Eight averaged. (Reproduced with permission
from Chapter 12, Manual of Remote Sensing, First
Edition, ©1975, Am. Soc. of Photogrammetry.)

than one image to work with. Even if multiple images are available, they must be precisely registered before averaging, or the processed image will be degraded.

2.4.2 Isolated Noise

Isolated bad pixels and lines in digital images can be caused by bit loss in data transmission, sudden detector saturation, or other intermittent electronic problems. The pixels affected by the noise usually have a zero gray level, indicating data loss, or a maximum gray level, indicating saturation. This type of noise, therefore, may be removed by comparing each pixel with its neighbors and deciding whether the pixel is bad or good based on its deviation from the neighboring pixels. This processing is analogous to editing techniques used in statistics to detect outliers or "wild-points" in a data set. If a pixel or line is bad it can be replaced by the nearest good pixel or line, or by a gray level interpolated between the neighbors. Care must be exercised, particularly in the removal of bad lines, that the neighboring pixels used in the comparison are in themselves good. An example *noise cleaning* algorithm is shown in Fig. 2-24. Note that slightly different algorithms are required to remove bad vertical or horizontal lines or isolated pixels.

A non-linear filtering technique, using the *Tukey median filter* (Frieden, 1976; Wecksung and Campbell, 1974) also has been applied to isolated noise removal. The procedure is similar to a low-pass filter but each pixel is replaced by the median gray level within the neighborhood, rather than the mean. The filter tends to remove isolated pixels or lines differing from their neighbors by a large amount, because they do not significantly affect the gray level median. The median filter does not work as well on an image with random noise at each pixel because there are no well-defined outliers relative to

good pixels. The performance of the median filter is compared
in Fig. 2-25 to the noise cleaning algorithm of Fig. 2-24 and
simple low-pass filtering. The median filter is clearly

(a) INDIVIDUAL PIXELS

GL_1	GL_4	GL_7
GL_2	GL_5	GL_8
GL_3	GL_6	GL_9

$AVE_1 = (GL_1 + GL_3 + GL_7 + GL_9)/4$
$AVE_2 = (GL_2 + GL_4 + GL_6 + GL_8)/4$
$DIFF = |AVE_1 - AVE_2|$
if: $|GL_5 - AVE_1|$ or $|GL_5 - AVE_2| > DIFF$,
then: $GL_5' = AVE_2$ otherwise: $GL_5' = GL_5$

(b) LINES

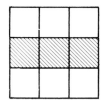

$AVE_1 = (GL_1 + GL_4 + GL_7)/3$
$AVE_2 = (GL_3 + GL_6 + GL_9)/3$
$DIFF = |AVE_1 - AVE_2|$
if: $|GL_5 - AVE_1|$ or $|GL_5 - AVE_2| > DIFF$,
then: $GL_5' = (GL_4 + GL_6)/2$ otherwise: $GL_5' = GL_5$

(c) COLUMNS

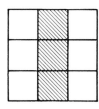

$AVE_1 = (GL_1 + GL_2 + GL_3)/3$
$AVE_2 = (GL_7 + GL_8 + GL_9)/3$
$DIFF = |AVE_1 - AVE_2|$
if: $|GL_5 - AVE_1|$ or $|GL_5 - AVE_2| > DIFF$,
then: $GL_5' = (GL_2 + GL_8)/2$ otherwise: $GL_5' = GL_5$

FIGURE 2-24. An example noise cleaning algorithm.

superior to the low-pass filter for removing bad lines; however, neither technique works as well as noise cleaning in terms of preserving the quality of noise-free areas.

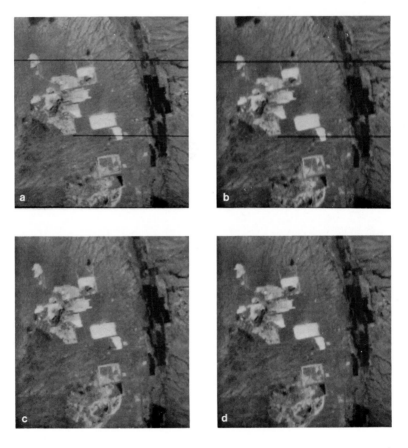

FIGURE 2-25.   Isolated noise suppression. (a) Image with noise.
               (b) Low-pass filter. (c) Median filter. (d) Noise
               cleaning.

2.4.3 Stationary Periodic Noise

Stationary periodic noise is a spurious, repetitive pattern that has consistent characteristics throughout an image. A typical source for such noise may be electronic interference from data transmission or reception equipment. The consistent periodicity of the noise is expressed as well-defined spikes in the Fourier transform of the noisy image (see Fig. 1-11). If the noise spikes are at a sufficient distance from the image spectrum (i.e., the noise is relatively high frequency), they may be removed by simply setting their amplitude to zero in the spectrum. The "cleaned" spectrum may then be inverse Fourier transformed to yield a noise-free image. An example of periodic noise removal by filtering is shown in Fig. 2-26. The spatial noise pattern is shown for illustration purposes only; it is not necessary to calculate this pattern to perform the processing. Spatial domain convolution filters may be designed that produce the same result and the choice of processing in the spatial or Fourier domain is primarily one of computational speed (see Sections 1.4.5 and 2.3).

If the frequency of the noise falls within the range of frequencies in the image spectrum, then real image structure will also be removed if the Fourier domain noise spikes are simply set to zero. This can be avoided to some extent by replacing the noise spikes in the Fourier domain by spectral values that have been *interpolated* from surrounding noise-free areas of the spectrum.

2.4.4 Non-Stationary Periodic Noise

If the noise is periodic but the amplitude, phase or frequency varies throughout the image, it can be called non-stationary. Linear spatial filtering is not suitable for this problem because the shift-invariant PSF used in linear filtering

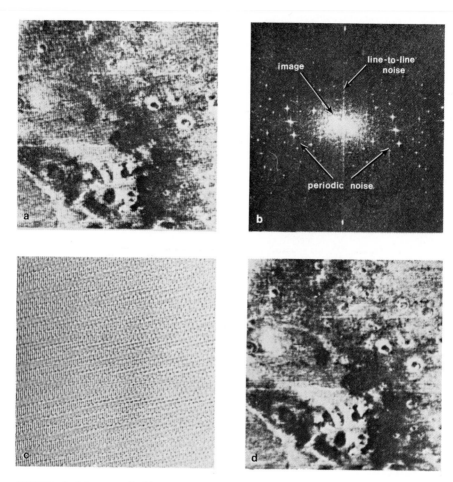

FIGURE 2-26. *Periodic, stationary noise suppression by spatial
filtering. (a) Image with noise. (b) Fourier spectrum of (a). (c) Periodic noise pattern. (d) Processed image. (Rindfleish et al., 1971)*

is independent of local image properties and therefore would either leave residual noise or, with complete noise removal, destroy a considerable amount of image structure in noise-free areas. An alternate approach suitable for single frequency, variable amplitude noise is to first obtain the stationary component of the spatial noise pattern by inverse Fourier transformation of the frequency domain noise spikes, as in Fig. 2-26. The local correlation between this noise pattern and the noisy image is then calculated for every image pixel (see Sections 1.4.4 and 2.5.1 for discussion of correlation). A fraction of the noise pattern amplitude, proportional to the magnitude of the local correlation at each pixel, is then subtracted from the noisy image. Thus, in areas where the noise component dominates, a larger fraction of the noise pattern is removed and conversely, in areas where the noise is weak and the image dominates, a proportionately smaller fraction of the noise pattern is removed. A simplified version of this procedure, implemented entirely in the spatial domain, was used by Chavez and Soderblum (1975) to remove the single frequency, variable amplitude noise in the Mariner 9 image in Fig. 2-27.

2.5 Spatial Registration

Precise image-to-image registration is necessary to form image mosaics, map temporal changes accurately, compare images from two different sensors, or combine multispectral images in a color composite. The accuracy to which images may be registered depends on many factors. Registration of multitemporal images, for example, is sometimes difficult because of the very temporal changes that we are interested in. Unless the images are one year apart, changing solar elevations will result in different shadow patterns that can produce misregistration if locations that are heavily shadowed are chosen for correlation. Lake

 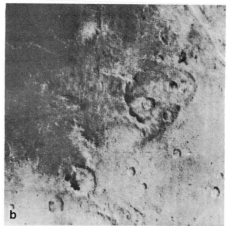

FIGURE 2-27. Periodic, non-stationary noise suppression by adaptive spatial filtering. (a) Image with noise. (b) Processed image. (Chavez and Soderblum, 1975; reproduced with permission of Proc. Am. Soc. of Photogrammetry Annual Convention, March 1975; ©1975, Am. Soc. of Photogrammetry)

boundaries can also change because of changes in water level, and are not reliable features for registration. Similarly, multispectral images can differ from each other considerably, making band-to-band registration difficult. Although most digital sensors are designed to provide multispectral data that is registered band-to-band, the increasing use of multi-source and multi-sensor data (see Sec. 3.7), has made the requirement for registering diverse sets of data commonplace.

2.5.1 Automatic Registration

Automated digital registration can be accomplished with some measure of similarity or dissimilarity between two images

that is calculated as a function of relative shift between them.
One similarity measure is the *correlation* between two overlapping image areas (Section 1.4.4).   If the two areas are
registered, the correlation value is nominally a maximum.
Because correlations are computationally expensive for large
areas, relatively small areas, distributed over the total overlapping region of the two images, are used (Fig. 2-28). These
areas are small enough that an x-y shift between the images is
sufficient for satisfactory correlation within each area.   The
differences between the optimum x-y shifts for the different
areas define any rotation, skew or other higher order misalign-
ments between the two full images.

To calculate the correlation, an N-by-N "window" area is
selected in one image and an M-by-M "search" area, with M
greater than N, is selected in the other image (Fig. 2-29). The
correlation is then performed between the window and search

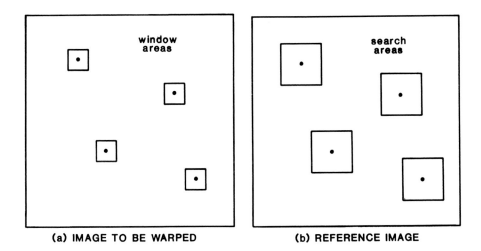

(a) IMAGE TO BE WARPED   (b) REFERENCE IMAGE

FIGURE 2-28. *Designation of areas to be correlated between*
            two images.

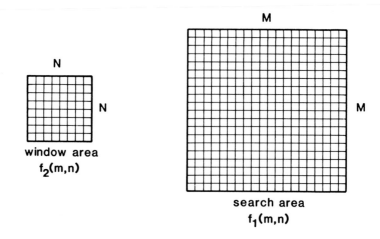

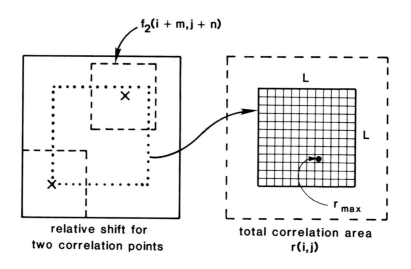

FIGURE 2-29. Correlation between window and search areas.

areas, over the central L-by-L region of the search area. Calculation of the correlation for all L^2 possible shift values requires on the order of N^2L^2 operations. Of course the window and search areas do not have to be square; the only requirement is that the search area be larger than the window area.

To prevent false correlation peaks arising from changes in the mean image gray level over the search area, the correlation defined by Eq. (1-20) is usually normalized in the following way (Pratt, 1978)

$$r(i,j) = \sum_{m=1}^{N} \sum_{n=1}^{N} f_1(m,n) f_2(i+m, j+n) / A_1 A_2 \qquad (2\text{-}2)$$

where

$$A_1 = \left[\sum_{m=1}^{N} \sum_{n=1}^{N} f_1^2(m,n) \right]^{1/2} \qquad (2\text{-}3)$$

and

$$A_2 = \left[\sum_{m=1}^{N} \sum_{n=1}^{N} f_2^2(i+m, j+n) \right]^{1/2} \qquad (2\text{-}4)$$

Correlation measures based on the difference between the two images may also be used for the same reason (Eq. [1-18]). However, although Eq. (1-20) may be computed via Fourier transforms as described in Sec. 1.4.4., Eqs. (1-18) and (2-2) can only be computed directly in the spatial domain. Techniques that significantly increase the speed of spatial domain correlations have been described by Barnea and Silverman (1972) and successfully applied to Landsat images by Bernstein (1976). These sequential similarity detection algorithms (SSDAs) use a small number of randomly located pixels within the window and search areas to quickly find the approximate point of registration, followed by full calculations using all the pixels in the window for precise registration.

In general, the SNR of a correlation calculation increases with the window size N. On the other hand, N must be small enough to preclude the presence of significant internal distortions within the window area. A window of about 51-by-51 pixels has been shown by Bonrud and Henrikson (1974) to yield *average* registration errors of about one-fourth to one-half pixel between two Landsat MSS images from different dates. Registration accuracy may vary considerably, however, from one image area to another because of differences in image contrast and detail. Edge enhancement filters (Sec. 2.3) may be used as a preprocessing technique to improve correlation (Pratt, 1978).

2.5.2 Manual Registration

Many remote sensing applications do not require the registration precision that is possible with automatic correlation. Furthermore, correlation algorithms sometimes produce spurious registration peaks, which may be detected easily by visual examination of the images, but nevertheless reduce the desirability of this approach. Numerical correlation cannot in most cases be used to register an image to a map. For these reasons the visual location of small ground features, called *control points*, in the two images or image and map to be registered is commonly used in practice. For carefully selected control points, registration accuracies within one pixel are possible. This approach is particularly efficient with the aid of a CRT image display and a movable cursor that can provide pixel coordinates in the displayed image. The primary limitation with this approach is the difficulty in finding satisfactory control points that are well-distributed across the image area to be registered.

2.6 Geometrical Manipulation

It is necessary to change the geometrical properties of a digital image, i.e., its spatial coordinate system, to correct for systematic pixel positional errors or perform image-to-image or image-to-map registration. An example of geometric correction for severe scanner-induced distortion is shown in Fig. 2-30. Once the required *coordinate transformation* is determined, either from mathematical models of the sensor scanning geometry or from image-to-image correlation data or control point measurements, it may be implemented by *interpolation* of the original image pixel grid. These two steps in the geometric manipulation of images, coordinate transformation and interpolation, are discussed in the following sections.

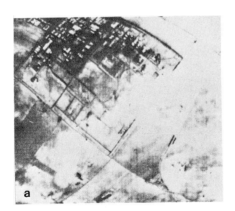 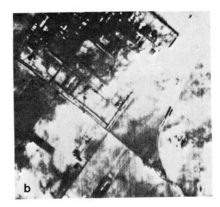

FIGURE 2-30. *Geometrical manipulation. (a) Distorted image. (b) Corrected image. (Wecksung and Campbell, 1974; ©1974 IEEE)*

2.6.1 Coordinate Transformation

Systematic geometric distortions are generated in Landsat MSS data by earth rotation during image scanning (this results in the familiar skew in the full scene MSS format), nonlinear scan mirror motion and the non-polar orbit of the satellite. Because these distortions are consistent over time, they can be corrected in a deterministic manner using orbital models and scanner calibration data (Anuta, 1973). These calibration procedures, however, do not correct for such factors as satellite roll, pitch and yaw, and changes in topographic elevation. These factors create generally unpredictable distortions that must be removed with the aid of control points and mathematical distortion models.

Corresponding control points, located in the image to be spatially warped and in the reference coordinate system (such as a map or another image) by either numerical or manual registration techniques, may be used to "tie" the two sets of data together. For example, the coordinate transformation within a *quadrilateral* defined by four control points may be modeled by a power series polynomial of the form

$$x = a_0 + a_1 x' + a_2 y' + a_3 x'y' + a_4 (x')^2 + a_5 (y')^2$$

$$y = b_0 + b_1 x' + b_2 y' + b_3 x'y' + b_4 (x')^2 + b_5 (y')^2 \qquad (2-5)$$

where the primed coordinates are those of the desired output image (reference system) and the unprimed coordinates are those of the original, distorted image. If the coefficients of the $(x')^2$ and $(y')^2$ terms are set to zero, the eight remaining constants can be found from the corner coordinates of the corresponding input and output quadrilaterals. The types of transformations possible in this case are shown in Fig. 2-31. By

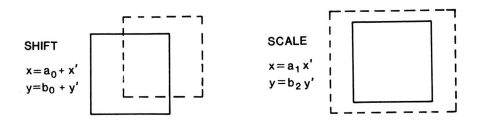

SHIFT

$x = a_0 + x'$
$y = b_0 + y'$

SCALE

$x = a_1 x'$
$y = b_2 y'$

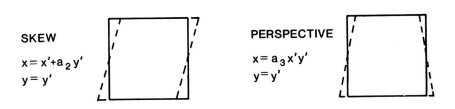

SKEW

$x = x' + a_2 y'$
$y = y'$

PERSPECTIVE

$x = a_3 x' y'$
$y = y'$

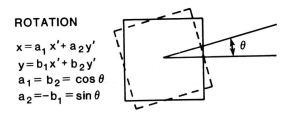

ROTATION

$x = a_1 x' + a_2 y'$
$y = b_1 x' + b_2 y'$
$a_1 = b_2 = \cos\theta$
$a_2 = -b_1 = \sin\theta$

FIGURE 2-31. Simple geometric transformations.

defining a net of contiguous quadrilaterals across the image (Fig. 2-32a) and a different polynomial transformation within each, a complex transformation can be accomplished with a simpler piecewise approximation. Retention of the squared terms in Eq. (2-5) necessitates a least squares fit of the polynomial to the control points at the corners of each quadrilateral; there is then some residual distortion at each control point. The technique of piecewise distortion modeling has been perfected by the Jet Propulsion Laboratory (Castleman, 1979) and applied extensively to severely distorted planetary spacecraft images.

An important question to consider before applying this technique is whether the actual distortion is modeled accurately by the piecewise polynomials. A dense net of control points will generally yield higher accuracy, but in remote sensing the analyst often is limited by the number and distribution of reliable control points. Also, the second order polynomial transformation of Eq. (2-5) is discontinuous across the boundary between two quadrilaterals, which can be a problem if there is severe distortion in the image. The use of contiguous *triangles* between three control points (Fig. 2-32a), with a corresponding deletion of the $(x')^2$, $(y')^2$, and $x'y'$ terms in the polynomial, eliminates the boundary discontinuities, but the transformation within each triangle is of a lower order. The global distortion is modeled in this case by a piecewise set of *planes*.

For relatively small portions of a Landsat scene, say 500-by-500 pixels, it is feasible to use a *single* polynomial model for image-to-image or image-to-map registration. This is because the Landsat data has a great deal of inherent geometric fidelity, and residual distortions, after preprocessing, are small. The procedure commonly followed is to locate at least six or seven control points distributed uniformly across the area of interest in the two sets of data. The coordinate

image coordinate system
(x,y)

reference coordinate system
(x',y')

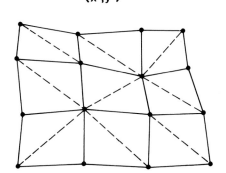

**(a) PIECEWISE APPROXIMATION WITH
QUADRILATERALS (——) OR TRIANGLES (— — —)**

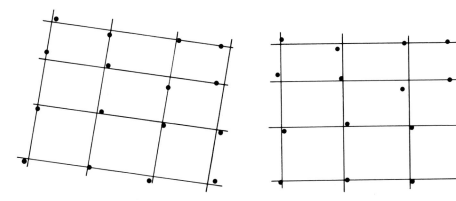

**(b) GLOBAL APPROXIMATION WITH LEAST SQUARES,
LINEAR POLYNOMIAL**

FIGURE 2-32. Distortion models using control points.

transformation is then modeled with a single low order, linear polynomial obtained by a least squares fit to the control points (Fig. 2-32b). This relatively simple *affine* transformation accounts for any shift, scale difference, rotation, or skew (Fig. 2-31) between the two sets of data and the least squares process mollifies any large errors in the location of individual control points. It is advisable, however, to edit the control points after a trial polynomial fit and to delete any points that are not approximated accurately by the polynomial.

2.6.2 Interpolation

For each column and row (x',y') in the processed image, the corresponding coordinate (x,y) in the original image must be calculated. Because the values of (x,y) calculated by Eq. (2-5) do not necessarily occur *exactly* at the coordinates of an original pixel, interpolation of the original image at (x,y) is necessary to calculate the pixel value to be inserted at (x',y') in the processed image. This situation is illustrated in Fig. 2-33.

The fastest scheme for calculating interpolated pixels for a geometrical transformation is *nearest-neighbor assignment* (sometimes called zero-order interpolation). For the value of each new pixel at (x',y') in the output image, the value of the original pixel *nearest* to (x,y) is selected. Because of this "round-off" property, geometric discontinuities on the order of plus or minus one-half a pixel interval are introduced in the processed image. This effect is negligible for most visual display applications, but may be important for subsequent numerical analyses. The significant advantage of nearest-neighbor assignment over other algorithms is that no calculations are required to derive the output pixel value, once the location (x,y) has been calculated (which must be done no matter what interpolation algorithm is used). Nearest-neighbor assignment

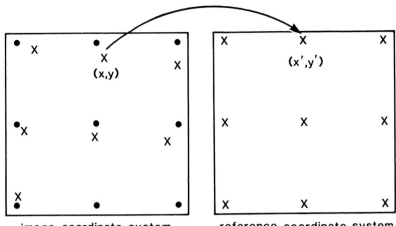

image coordinate system   reference coordinate system

● original pixels
X interpolated pixels

FIGURE 2-33. Implementation of a geometrical transformation.

is equivalent to convolving the input image with a uniform PSF that is one sample interval wide as shown in Fig. 2-34a. Unlike spatial filtering, however, output pixel values are calculated at locations *between* the original pixels.

A geometrically smoother interpolated image is obtained with *bilinear* (first-order) interpolation. This algorithm uses the four input pixels surrounding the point (x,y) to estimate the output pixel. Bilinear interpolation is commonly implemented by convolving the input image separately in the x and y directions with a triangle weighting function (Fig. 2-34a). The difference between the nearest—neighbor and bilinear algorithms may be appreciated by the comparison in Fig. 2-35. The image has been enlarged digitally by the two methods to illustrate the

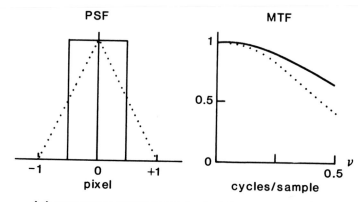

(a) NEAREST NEIGHBOR (—) AND BILINEAR (···)

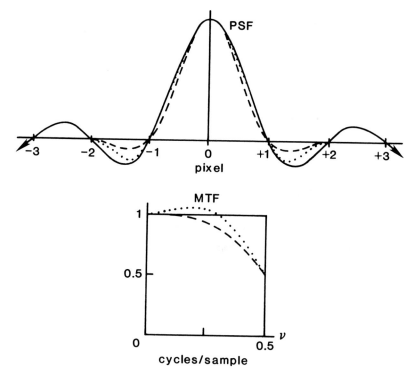

(b) SINC (—), STANDARD CUBIC (···) AND IMPROVED CUBIC (– – –)

FIGURE 2-34.   The PSF and MTF of several interpolators.

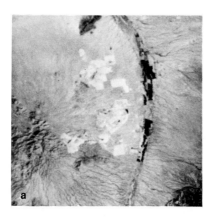

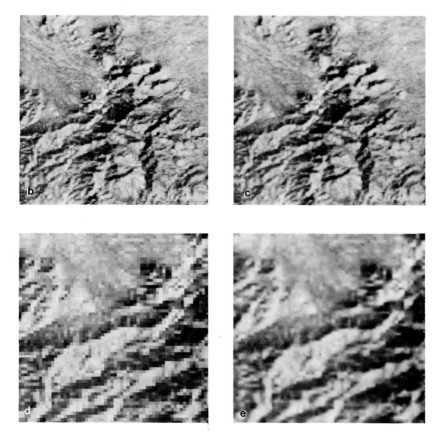

*FIGURE 2-35.   Image magnification by interpolation. (a) Original.
(b, c) 4×: nearest neighbor and bilinear. (d, e)
8×: nearest neighbor and bilinear.*

smoother, more continuous nature of bilinear interpolation. The bilinear algorithm, however, requires nine calculations (add, subtract, or multiply) for each output pixel (Castleman, 1979) and is therefore considerably slower than nearest-neighbor processing.

The smoothing incurred with bilinear interpolation may be avoided with *cubic* (second-order) interpolation, at the expense of more computation. The cubic interpolating function is a piecewise cubic polynomial (Bernstein, 1976; Fig. 2-34b) that approximates the theoretically ideal interpolation function for imagery, the sinc ($\sin[\pi x]/\pi x$) function (Pratt, 1978). The sinc function is normally not used for image interpolation because a large pixel neighborhood is required for accurate results (Goetz et al, 1975). The cubic interpolator yields results approximating those from a sinc function and requires only a 4-by-4 pixel array in the input image for interpolation.

The effect of the nearest-neighbor, linear, and standard cubic interpolators on Landsat imagery can be seen in Fig. 2-36. The unprocessed Landsat image contains the original, non-unity sampling aspect ratio, which is normalized in the interpolated images of Fig. 2-36. Cubic interpolation generally does not yield a pronounced visual improvement over bilinear interpolation and requires about twice as much computation time on a large IBM computer according to Bernstein (1975). For applications in which the display process itself introduces a considerable amount of resolution degradation (for example, image maps in the form of a lithographic print) nearest-neighbor assignment may produce a satisfactory product with a large savings in computer processing time compared to the other interpolation functions.

A drawback of standard cubic interpolation is that it produces some gray level overshoot on either side of sharp edges. Although this characteristic contributes to the visual sharpness of the processed image, it is undesirable for further numerical analyses where radiometric accuracy is of prime importance.   It has been shown recently (Keys, 1981; Park and Schowengerdt, 1983) that another member of the cubic family of functions is preferable in many ways to the standard cubic.   The two interpolation functions are shown in Fig. 2-34b and compared by interpolation across an edge in Fig. 2-37.   Gray level overshoot is greatly reduced with the improved cubic interpolator.

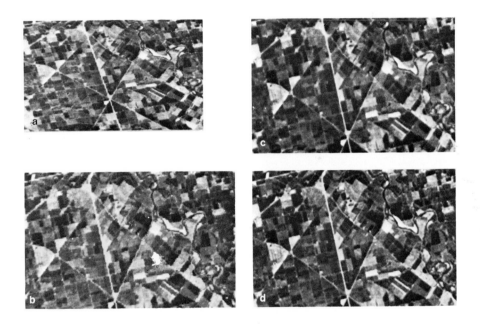

FIGURE 2-36.   *Image interpolation with three different interpolators. (a) Original. (b) Nearest-neighbor. (c) bilinear. (d) Cubic. (Bernstein and Ferneyhough, 1975; reproduced with permission from Photogrammetric Eng. and Remote Sensing, December 1975; ©1975, Am. Soc. of Photogrammetry)*

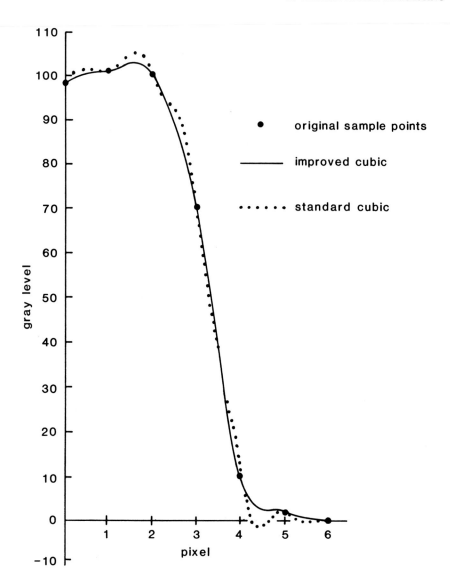

FIGURE 2-37. *Interpolated edge profile using two different cubic interpolators.*

2.7 Color Processing

All of the processing examples presented thus far have been black and white (monochrome) images. The use of color in display and enhancement of remote sensing images, however, is an important aspect of image processing, particularly with the increasing use of interactive color CRT displays. Color may be used simply for display of multispectral images or may be manipulated directly by processing techniques to enhance visual information extraction from the images. In this section we describe some relatively simple heuristic techniques for numerical color manipulation; theories that model the visual perception of color can be quite complex (Faugeras, 1979) and will not be discussed here.

2.7.1 Color Composites

All color CRT systems for display of digital images utilize an *additive* color composite system with three *primary colors*: red, green, and blue (RGB). Figure 2-38 depicts a simplified schematic of such a display system. The LUTs are used to control the contrast in each channel (Appendix C). Three bands of a multispectral image are typically stored in the three refresh memories and displayed in color composite form with one of the multispectral bands displayed in red, one in green, and one in blue. If Landsat MSS bands 7, 5, and 4 are displayed in red, green, and blue, respectively, a standard *false color infrared* image is obtained (Plate la). This is a commonly used color combination for Landsat MSS composites because the image is similar to color infrared photography, particularly in the red rendition of vegetation (lower right corner of Plate la). With digital multispectral images and easily controlled CRT displays, however, the color assignments for the component bands are *arbitrary*, as shown in Plate lb. This is especially evident

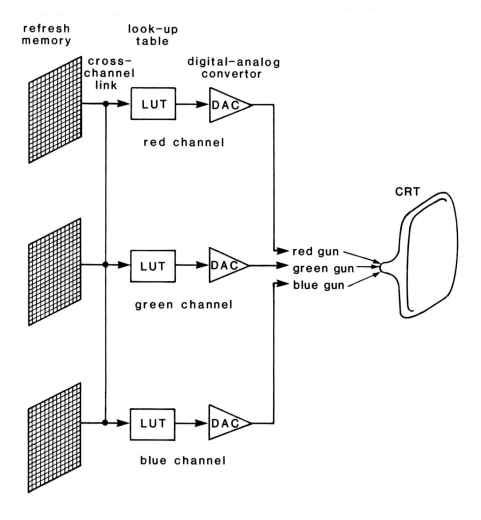

FIGURE 2-38. Simplified schematic of a RGB color CRT image display.

if the image spectral bands do not correspond to the spectral response ranges of color photographic emulsions. For example, thermal and visible band images may be combined in color, or spatially-registered non-image and image or image-derived data may be composited in a single color representation (Plate 2).

The gray levels of each component in an RGB display constitute the orthogonal axes of a three-dimensional color space; the maximum possible gray level in each channel of the display defines the RGB *color cube* (Fig. 2-39). Every pixel composited in this display system may be represented by a vector that lies somewhere within the color cube. For a display with 8 bits/pixel/band, a total of 256^3 colors are ideally possible. Limitations in CRT and visual color response, however, reduce the number of colors that can be visually discriminated. The line from the origin of the color cube to the opposite corner is

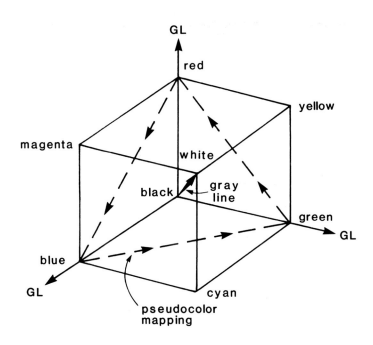

FIGURE 2-39.   The RGB color cube.

known as the *gray line* because pixel vectors that lie on this
line have equal components in red, green and blue.

2.7.2 Pseudocolor

Color may be used very effectively to enhance small gray
level differences in a monochrome image. To create a pseudocolored (sometimes called color-sliced or color-coded) image,
each gray level is simply displayed as a unique color. There
are many ways to logically assign colors to gray levels; one
approach is shown in Fig. 2-40. The locus of pixel vectors in
RGB space for this particular transformation is shown in Fig.
2-39. The pseudocolor mapping may be implemented by distribut-
ing the contents of the refresh memory that contains the monochrome image through the three color channels (see cross-channel
link in Fig. 2-38). The three LUTs may then be programmed as in
Fig. 2-40a.

An example of image pseudocoloring is shown in Plate 3.
The gray scale added to the image provides an important calibration reference for the pseudocolor coding. Note how adjoining
gray levels are easily distinguished with the use of color. The
pseudocolor algorithm used here is different than that described
above, but produces similar results.

2.7.3 Color Transformations

To describe the color properties of an object in an image,
we do not normally use the proportions of red, green, and blue
components, but rather terms such as "intensity", "hue" and
"saturation" that describe the subjective sensations of
"brightness", "color" and "color purity", respectively. Like-
wise, it is often easier to anticipate the visual results of
intensity or hue manipulation in a color display than it is for
red, green, and blue manipulation. A transformation of the RGB
components into intensity, hue and saturation (IHS) components

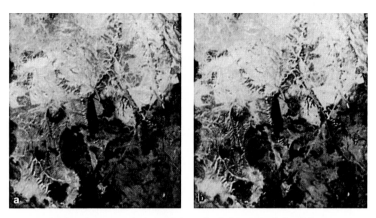

PLATE 1. Landsat MSS color composites. (a) False color infrared. Red, band 7; green, band 5; blue, band 4. (b) False color. Red, band 4; green, band 5; blue, band 7.

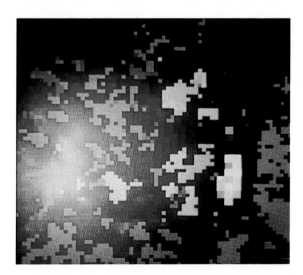

PLATE 2. Use of color to display simultaneously three different types of data. Elevation, red. Population density, green. Landsat MSS classification: mining and waste, light blue; bare soil, dark blue. (Glass and Schowengerdt, 1983)

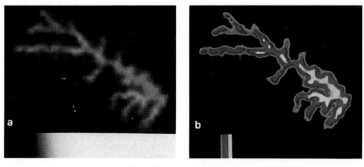

PLATE 3. Pseudocolor processing of a Heat Capacity Mapping Mission (HCMM) thermal image of Lake Anna, Virginia. (a) Gray level image. (b) Pseudocolor image of selected gray level range.

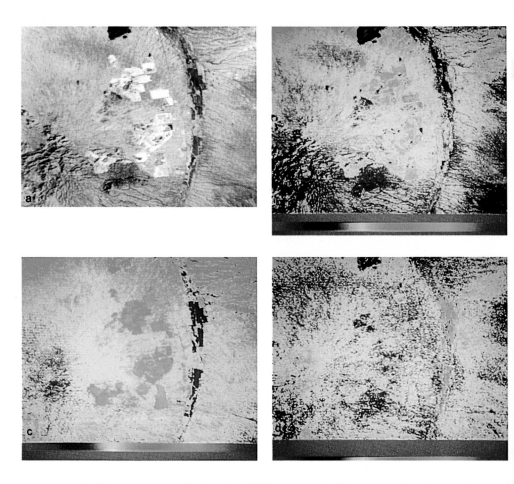

PLATE **4.** Color components of a Landsat MSS image.   (a) False color infrared composite.
(b) Pseudocolored intensity component.   (c) Pseudocolored hue component.   (d) Pseudocolored
saturation component.   (Image processing courtesy of Johann Pramberger)

before processing may therefore provide more control over color enhancement. The processed images are then converted back to RGB for display (Fig. 2-41). This type of processing has been implemented in hardware in one particular display system (Buchanan, 1979).

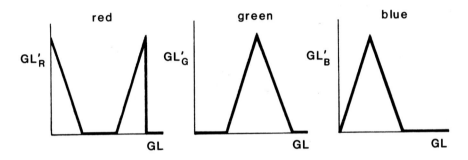

(a) GRAY LEVEL TRANSFORMATIONS
IN EACH DISPLAY CHANNEL

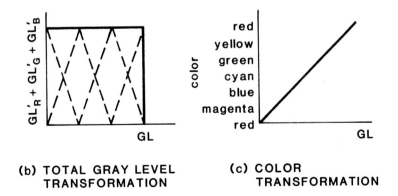

(b) TOTAL GRAY LEVEL
TRANSFORMATION

(c) COLOR
TRANSFORMATION

FIGURE 2-40. A pseudocolor transformation.

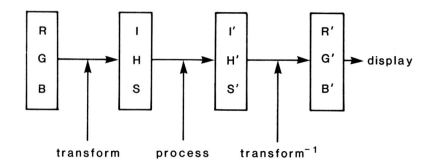

FIGURE 2-41. Image processing in color space.

We will describe one particular RGB to IHS transformation to illustrate the concepts involved. This transformation, the *hexcone model*, is heuristic and not based on any particular color theory, but it is representative of most algorithms used in color image processing and color graphics (Smith, 1978).

Imagine the *projection* of an RGB color subcube, with the vertex farthest from the origin defined by a point on the gray line, onto a plane perpendicular to the gray line at that point.   Moving the vertex from black to white, a series of hexagons of increasing size results as shown in Fig. 2-42a. The hexagon at black degenerates to a point; the hexagon at white is the largest. This series of hexagons define a solid called the *hexcone*. The distance from black along the gray line defines the *intensity* of each hexagonal projection.[1] For a pixel with a

---

[1]Smith (1978) defined an alternate quantity, *value*, given by the maximum of R, G, and B. Value is more closely related to artist's terminology for describing color. The distinction between value and intensity is important, but not crucial for our discussion.

given intensity, the color components, hue and saturation, are defined geometrically in the appropriate hexagon (Fig. 2-42b). The *hue* of a point in each hexagon is determined by the angle around the hexagon and the *saturation* is determined by the distance of the point from the center, i.e., the gray point of the hexagon. Points further from the center represent purer colors than those closer to the gray point. The use of simple linear distances for defining hue and saturation make the hexcone algorithm more efficient than similar transformations involving trigonometric functions.

An example of an image transformed by the hexcone model is shown in Plate 4. The IHS components are each pseudocolored to more clearly display their characteristics. The intensity component includes the effects of topographic shading in the mountains (lower left), whereas hue and saturation are nearly independent of this factor. The hue component clearly separates vegetated and non-vegetated areas, and the saturation component indicates that the densely vegetated agricultural fields (right center) have a relatively pure color.

The IHS components may be used as an intermediate step for image enhancement. For example, a contrast stretch can be applied to the intensity component only, and will *not affect* the hue and saturation of pixels in the processed image (R'G'B' in Fig. 2-41). There will be no numerical color shift in the enhanced image, a characteristic that is not generally true if the R, G, and B components are contrast stretched directly. The IHS transformation is also useful for displaying diverse, but spatially-registered images. For example, a high resolution visible band image may be displayed as the intensity component and a lower resolution thermal band image as the hue component (Haydn et al, 1982). The resulting R'G'B' image contains the detail structure of the scene expressed as intensity, with the thermal structure superimposed as pure color variations.

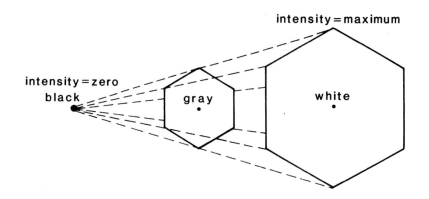

(a) GENERATION OF THE HEXCONE

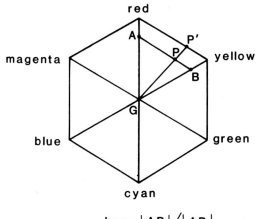

$$\text{hue} = |AP| / |AB|$$
$$\text{saturation} = |GP| / |GP'|$$

(b) DEFINITION OF COLOR COMPONENTS
FOR A PIXEL VECTOR WITH NON-ZERO
INTENSITY AT POINT P

FIGURE 2-42. The hexcone color model.

References

Andrews, Harry C. and B. R. Hunt, Digital Image Restoration, Englewood Cliffs, New Jersey, Prentice-Hall, Inc., 1977.

Anuta, Paul E., "Geometric Correction of ERTS-1 Digital Multi-spectral Scanner Data," LARS Information Note 103073, Purdue University, Laboratory for Applications of Remote Sensing, 1973.

Arguello, R.J., "Encoding, Transmission and Decoding of Sampled Images," A Symposium on Sampled Images, Perkin-Elmer Corp. publication, 1971, pp. 3-1 - 3-21.

Barnea, D.I. and H.F. Silverman, "A Class of Algorithms for Fast Digital Image Registration," IEEE Trans. Computers, Vol. C-21, February 1972, pp. 179-186.

Bernstein, R., "Digital Image Processing of Earth Observation Sensor Data," IBM Journal of Research and Development, Vol. 20, No. 1, January 1976, pp. 40-57.

Bernstein, R. and D.G. Ferneyhough, Jr., "Digital Image Processing," Photogrammetric Engineering and Remote Sensing, Vol. XLI, No. 12, December 1975, pp. 1465-1476.

Biberman, L.M., ed., Perception of Displayed Information, New York, Plenum Press, 1973.

Blackwell, R.J. and D.H. Boland, "The Trophic Classification of Lakes using ERTS Multispectral Scanner Data," Proc. Am. Soc. of Photogrammetry, Am. Soc. of Photogrammetry, Falls Church, Va., March 1975, pp. 393-413.

Bonrud, L.O. and P.J. Henrikson, "Digital Registration of ERTS-1 Imagery," IEEE Conference on Decision and Control, Phoenix, Arizona, November 1974.

Brigham, E. Oran, The Fast Fourier Transform, Englewood Cliffs, New Jersey, Prentice-Hall, 1974, 252 pp.

Buchanan, Michael D., "Effective Utilization of Color in Multi-dimensional Data Presentations," Proc. Soc. of Photo-optical Instrumentation Engineers, Vol. 199, Advances in Display Technology, 1979, pp. 9-18.

Castleman, Kenneth R., Digital Image Processing, Englewood
    Cliffs, New Jersey, Prentice-Hall, Inc., 1979, 429 pp.

Chavez, P.S. Jr. and L.A. Soderblum, "Simple High-Speed Digital
    Image Processing to Remove Quasi-Coherent Noise
    Patterns," Proc. Am. Soc. of Photogrammetry, Falls
    Church, Va., March 1975, pp. 595-600.

Davis, Larry S., "A Survey of Edge Detection Techniques,"
    Computer Graphics and Image Processing, Vol. 4, 1975,
    pp. 248-270.

Fahnestock, James D. and Robert A. Schowengerdt, "Spatially-
    Variant Contrast Enhancement Using Local Range
    Modification," Optical Engineering, Vol. 22, No. 3,
    May-June 1983.

Faugeras, Olivier D., "Digital Color Image Processing Within the
    Framework of a Human Visual Model," IEEE Trans. on
    Acoustics, Speech, and Signal Processing, Vol. 27, No.
    4, 1979, pp. 380-393.

Frieden, B.R., "A New Restoring Algorithm for Preferential
    Enhancement of Edge Gradients," J. Opt. Soc. Am., Vol.
    66, No. 3, March 1975, pp. 280-283.

Glass, C.E. and R.A. Schowengerdt, "Hazard and Risk Mapping of
    Mined Lands Using Satellite Imagery and Collateral
    Data," The Bulletin of the Association of Engineering
    Geologists, Vol. 20, No. 2, April 1983.

Goetz, A.F.H., F.C. Billingsley, A.R. Gillespie, M.J. Abrams,
    R.L. Squires, E.M. Shoemaker, I. Luchitta and D.P.
    Elston, "Application of ERTS Images and Image Proces-
    sing to Regional Geologic Problems and Geologic Mapping
    in Northern Arizona," Jet Propulsion Laboratory
    Technical Report 32-1597 prepared for NASA Contract 7-
    100, California Inst. of Technology, May 1975, Chapter
    III and Appendix B.

Gonzalez, Rafael C. and Paul Wintz, Digital Image Processing,
    Reading, Mass., Addison-Wesley, 1977, 431 pp.

Goodman, Joseph W., Introduction to Fourier Optics, New York,
    McGraw-Hill, 1968, 287 pp.

Haydn, Rupert, George W. Dalke, and Jochen Henkel, "Application
    of the IHS Color Transform to the Processing of Multi-
    sensor Data and Image Enhancement," Proc. International
    Symposium on Remote Sensing of Environment - First

Thematic Conference: Remote Sensing of Arid and Semi-arid Lands, Cairo, Egypt, January 1982, pp. 599-616.

Holkenbrink, Patrick F., Manual on Characteristics of Landsat Computer-Compatible Tapes Produced by the EROS Data Center Digital Image Processing System, U.S. Geological Survey, 1978, 70 pp.

Keys, R.G., "Cubic Convolution Interpolation for Digital Image Processing," IEEE Trans. on Acoustics, Speech, and Signal Processing, Vol. 29, No. 6, 1981, pp. 1153-1160.

Lee, Jong-Sen, "Digital Image Enhancement and Noise Filtering by Use of Local Statistics," IEEE Trans. on Pattern Analysis and Machine Intelligence, Vol. PAMI-2, No. 2, March 1980, pp. 165-168.

McDonnell, M.J., "Box-filtering Techniques," Computer Graphics and Image Processing, Vol. 17, No. 1, September 1981, pp. 65-70.

O'Handley, D.A. and W.B. Green, "Recent Developments in Digital Image Processing at the Image Processing Laboratory at the Jet Propulsion Laboratory," Proc. IEEE, Vol. 60, No. 7, July 1972, pp. 821-828.

Park, Stephen K. and Robert A. Schowengerdt, "Image Reconstruction by Parametric Cubic Convolution," Computer Vision, Graphics, and Image Processing, Vol. 20, No. 3, September 1983.

Pratt, William K., Digital Image Processing, New York, John Wiley and Sons, 1978, 750 pp.

Pun, T., "Entropic Thresholding, A New Approach," Computer Graphics and Image Processing, Vol. 16, Academic Press, 1981, pp. 210-239.

Reader, Clifford and Larry Hubble, "Trends in Image Display Systems," Proc. IEEE, Vol. 69, No. 5, May 1981, pp. 606-614.

Rindfleish, T.C., J.A. Dunne, H.J. Frieden, W.D. Stromberg, and R.M. Ruiz, "Digital Processing of the Mariner 6 and 7 Pictures," J. Geophysical Research, Vol. 76, No. 2, January 10, 1971, pp. 394-417.

Robinson, G.S., "Detection and Coding of Edges Using Directional Masks," Optical Engineering, Vol. 16, No. 6, November-December 1977.

Schafer, David H. and James R. Fischer, "Beyond the Super-
    computer," IEEE Spectrum, Vol. 19, No. 3, March 1982,
    pp. 32-37.

Schowengerdt, R., L. Babcock, L. Ethridge, and C. Glass, "Corre-
    lation of Geologic Structure Inferred from Computer-
    enhanced Landsat Imagery with Underground Water
    Supplies in Arizona," Proc. of the Fifth Annual William
    T. Pecora Memorial Symposium on Satellite Hydrology,
    American Water Resources Association, 1981.

Sellner, H.R., "Transfer Function Compensation of Sampled
    Imagery," A Symposium on Sampled Images, Perkin-Elmer
    Corp. publication, 1971, pp. 4-1 - 4-14.

Smith, Alvy Ray, "Color Gamut Transform Pairs," Proc. of the
    ACM-SIGGRAPH conference, Vol. 12, No. 3, 1978, pp. 12-
    19.

Steiner, Dieter and Anthony E. Salerno, Coauthors-Editors,
    "Remote Sensor Data Systems, Processing, and Manage-
    ment," pp. 611-803 in Manual of Remote Sensing, First
    Edition, Robert G. Reeves, ed., Am. Soc. of Photogram-
    metry, 1975, 2144 pp.

Wallis, R., "An Approach to the Space Variant Restoration and
    Enhancement of Images," in Proc. Seminar on Current
    Mathematical Problems in Image Science, Naval Post-
    graduate School, Monterey, Calif., Nov. 1976.

Wecksung, G.W. and K. Campbell, "Digital Image Processing at
    EG&G," Computer, Vol. 7, No. 5, May 1974, pp. 63-71.

Zweig, H., A. Silverstri, P. Hu, and E. Barrett, "Experiments in
    Digital Restoration of Defocused Grainy Photographs by
    Noise Cheating and Fourier Techniques," Proc. Soc. of
    Photo-optical Instrumentation Engineers, Vol. 74, Image
    Processing, 1976, pp. 10-16.

CHAPTER 3

Digital Image Classification

3.1 Introduction

In Chapter 2 we described image processing techniques that assist the analyst in the *qualitative*, i.e., visual, interpretation of images. In this chapter techniques are described that assist the analyst in the *quantitative* interpretation of images. In spite of this well-defined functional difference between image enhancement and image classification, both approaches to extracting information from images can sometimes benefit from the same techniques, for example in the preprocessing of imagery for improving classification.

Multispectral classification is emphasized in this chapter because it is, at the present time, the most common approach to computer-assisted mapping from remote sensing images, and lends itself well to discussion of the basic concepts that apply to all types of classification. It is important at this point, however, to make a few appropriate comments about multispectral classification. First, it is fundamental that we are attempting to objectively map areas on the ground that have similar spectral reflectance characteristics. The resulting labels assigned to the image pixels therefore represent *spectral classes* that may or may not correspond to the classes of ground objects that we are ultimately interested in mapping. A good example of such a situation is the mapping of urban land use from multispectral imagery. Urban land use classes, such as "urban residential" or "light industrial," are seldom character- ized by a single spectral signature, because they are composed of several land *cover* types (such as vegetation, pavement, and buildings), each having a different spectral signature. Within an area of particular land use, therefore, several spectral classes occur, resulting in an overall heterogeneous spectral signature for the land use class. Thus, we must look for more complex relationships between the physical measurements, the pixel-by-pixel multispectral image, and the map classes of interest. Incorporation of spatial texture in classification (Sec. 3.4.5) is one approach to this higher level of information extraction.

Second, manually-produced maps are the result of a long, often complex process that utilizes many sources of information. The conventional tools used to produce a map range from the strictly quantitative techniques of photogrammetry and geodesy, to the less quantitative techniques of photointerpretation and field class descriptions, to the subjective and artistic techniques of map "generalization" and visual extrapolation of discrete spatial data points. In the photointerpretation process alone, the analyst draws on many sources of information other than the image, including previous experience and knowledge of the area, and uses complex deductive logic to assimilate all of these components. In this context, image classification, and multispectral classification in particular, represent relatively narrow, albeit quantitative, approaches to mapping. It is therefore often appropriate to consider classification maps to be only one component of an ensemble that leads to a final map.

The final output of the classification process is a type of digital image, specifically a *map* of the classified pixels. For display, the class at each pixel may be coded by character or graphical symbols (Fig. 3-1) or by color (Plate 5). The classification process compresses the image data by reducing the large number of gray levels in each of several spectral bands into a

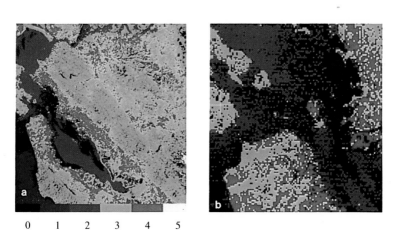

0   1   2   3   4   5

PLATE 5. Multispectral classification of the image in Fig. 1-6. (a) Full image. (b) Enlarged portion of (a). Spectral classes: 0, threshold; 1, clear water; 2, turbid water; 3, vegetation; 4, urban; 5, cloud.

few number of classes in a single image. In that sense, classification causes a *loss* of numerical information from the original image data.

3.2 Multispectral Classification

Most of the image processing techniques discussed in
Chapter 2 do not make explicit use of the spectral information
contained in a scene. In remote sensing, however, this information may provide the most significant clues about what is on the
ground. The spectral information in a scene can be recorded
with *multispectral* images, i.e., a set of images of the same
scene, each acquired through a different spectral filter (Sec.
1.2.2). Each pixel in a multispectral image has the spatial
coordinates x and y and the spectral coordinate λ (wavelength)
(Fig. 3-2). The spectral dimension is quantized, however, into
a few discrete spectral bands. Each pixel in any one spectral
band has a spatially coincident pixel in all the other bands.
For a K-band image, there are K gray levels associated with each
pixel, one for each spectral band. The K gray levels define a
K-dimensional *spectral measurement space* in which each pixel is
represented by a vector (Sec. 1.5.3).

3.2.1 Spectral Signatures

Relying on the assumption that different surface materials
have different spectral reflectance (in the visible and micro-
wave regions) or thermal emission characteristics, multispectral
classification logically partitions the large spectral measurement space (256^K possible pixel vectors for an image with 8
bits/pixel/band and K bands) into relatively few regions, each
representing a different type of surface material. Suppose, for
example, that we want to classify a multispectral image into the
very general classes of soil, water, and vegetation. Figure 3-3
shows typical spectral reflectance curves for these materials in

the visible and near infrared spectral regions. The set of discrete spectral radiance measurements provided by the broad spectral bands of the sensor define the *spectral signature* of each class, as modified by the atmosphere between the sensor and the ground. The spectral signature is a K-dimensional vector whose coordinates are the measured radiance in each spectral band. Figure 3-4 is a two-band plot of the spectral signatures of the classes in Fig. 3-3 showing that the three classes can be separated readily by partitions in only one of the spectral bands. The classification decision boundaries can therefore be simply gray level thresholds in band 7.

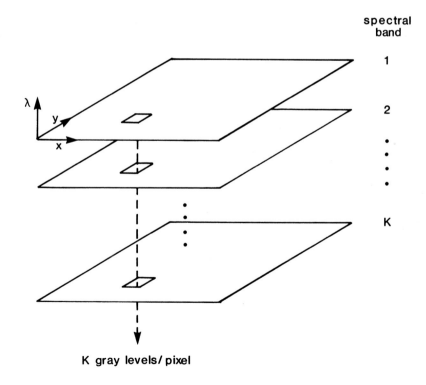

FIGURE 3-2. *The three coordinates of a multispectral image.*

In reality, the spectral radiance of a given surface material is not characterized by a single, deterministic curve, but by a *family* of curves with a range of variability (Fig. 3-5) caused by many natural factors (Sec. 3.2.2). Sensor properties, such as detector noise, can cause additional variation in the measured radiance. The pixel spectral measurements, therefore, form a *cluster* of vectors for each class (Fig. 3-6). Separation of the classes is now more difficult because, although water can still be distinguished from the other two classes, soil and vegetation overlap (a common situation) and require a compromise partition. To a large extent, our ability to perform an accurate

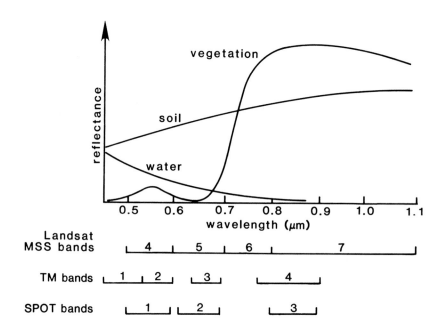

FIGURE 3-3. Generalized spectral reflectance curves for water, soil and vegetation.

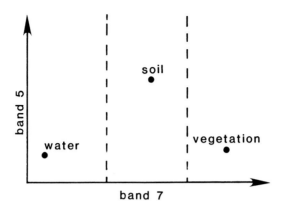

FIGURE 3-4. Two-band signatures for each class in Fig. 3-3.

classification of a given multispectral image is determined by the extent of overlap between class signatures. As discussed in Chapter 1, one compromise that can be achieved is minimization of the average error in the classification.

In general, the separation of all classes requires more than two spectral bands. Because the clusters occur in K-dimensional space, the class partitions are surfaces in K dimensions. At this point, we introduce the general term *feature* to describe each dimension of this K-dimensional space. As will be seen later, spectral bands are not the only possible components of this space; other image-derived properties may be useful, for example spatial texture or spectral ratios. The word feature accommodates this broader scope.

The distribution of actual data in feature space is usually difficult to visualize if K is greater than 2. The number of classes is typically six or more, which further confuses the display of class clusters in more than two dimensions.

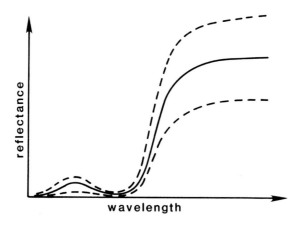

FIGURE 3-5. Statistical variation of reflectance for vegetation.

Consequently, the projection of the K-dimensional data onto one or two dimensions is commonly used for presentation purposes or manual analysis (Fig. 3-7). The projection onto one dimension simply results in the histogram of the data in that dimension. The projection onto two dimensions is called a *scattergram*, correlation plot, or two-dimensional histogram.

A final point can be illustrated with the two-dimensional classification example of this section. Suppose there is a group of pixel vectors at some point A in feature space (Fig. 3-6). These pixels would be classified as water with the class partitions as shown, although they may be a completely different material. Such points are sometimes referred to as *outliers* because they are unlike any of the specified classes. There is a need, therefore, to not only separate the specified classes from each other, but also from unknown or unwanted classes.

This is achieved with an operation known as thresholding (Sec. 3.6.1).

3.2.2 Natural Variables

Innumerable factors can cause data variability within class clusters and overlap between clusters. Among the most prominent are atmospheric scattering, topography, sun and view angles, class mixture, and within-class reflectance variability. A brief description of these factors is presented here to convey an appreciation of the many elements, in addition to surface spectral reflectance, that contribute to the image recorded by a satellite or aircraft sensor.

Atmospheric scattering

Atmospheric scattering adds a spectrally-dependent radiance component to the multispectral image, shifting gray levels by an amount that is virtually zero in the near-infrared and red regions of the spectrum and increases toward the blue-green

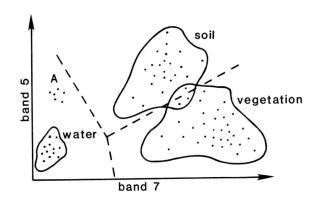

FIGURE 3-6. Typical two-band signatures for real data.

(a) PROJECTION ONTO TWO DIFFERENT PLANES

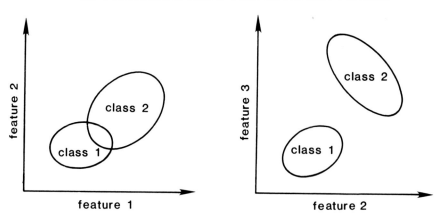

(b) PROJECTION ONTO TWO DIFFERENT AXES

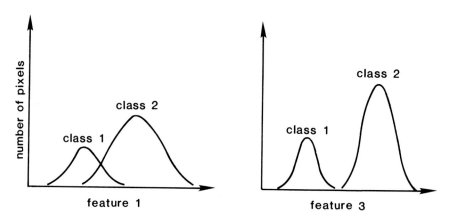

FIGURE 3-7. *Dimensionality reduction in the display of*
multidimensional data.

region (Fig. 3-8a). The *relative positions* of class clusters remain unchanged because the radiation scattered back to the sensor *before* reaching the ground (the major scattering component) is independent of the surface material. Classification is therefore not affected if training signatures (Sec. 3.3) are developed from the same image. If, however, the amount of atmospheric scattering varies from region-to-region in the image, or from date-to-date in a multitemporal image set, the classification may be seriously affected. Techniques for atmospheric scattering correction are described later (Sec. 3.4.1).

Topography

It is a familiar fact to photointerpreters that terrain slope and aspect affect scene radiance. As a simple illustration, imagine a hill covered with a single type of vegetation; the side facing the sun will have the same intrinsic color as

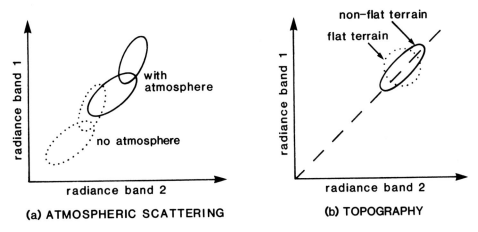

(a) ATMOSPHERIC SCATTERING   (b) TOPOGRAPHY

FIGURE 3-8. Two factors that influence spectral signatures.

the side facing away from the sun, but will appear brighter, thus producing an elongation in the vegetation spectral signature (Fig. 3-8b).

Because the cluster elongation caused by surface topography is *independent* of spectral band, it is aligned along a straight line through the origin of the spectral axes and can therefore be compensated for by computing a ratio between spectral bands (Sec. 3.4.2). The actual situation, however, is complicated by the atmosphere. The radiance of surface elements that receive *less* solar irradiance contains a relatively *greater* component from downward atmospheric scattering. This light, which is predominately blue, is in turn reflected by the surface towards the sensor. In the extreme, surface elements that are completely in shadow are not black and void of radiance variations, but produce some reflectance information that is weighted heavily towards the blue region of the spectrum. Multispectral atmospheric corrections are therefore necessary *before* the calculation of spectral ratios, if the latter are to be most effective.

Sun and view angles

Most materials have different reflectance properties in different directions. Thus the same material may have a different radiance and/or color in different portions of a scene because it is viewed from different angles, or its appearance may vary from scene-to-scene because the sun angle changes. An additional factor affecting scene radiance is terrain surface relief. Even if a given surface material has equal radiance when viewed from different angles, i.e. it is *Lambertian*, its radiance will depend on the cosine of the angle between the surface normal vector and the solar vector, as described in the previous section.

Class Mixtures

Many pixels in an image contain a mixture of surface cover classes, for example grass and underlying soil (Tucker and Miller, 1977). As the mixture proportions change from pixel-to-pixel, the spectral vector changes. Thus an overlap is created between the individual signatures of the mixed classes. This mixing is particularly troublesome in some applications, such as the mapping of soils or vegetation in sparsely vegetated arid areas. There has been some research into techniques for estimating class proportions in mixture pixels (sometimes called "mixels"!) by Horwitz, et al (1971), Salvato (1973) and others, but these techniques are not widely used.

One source of mixture pixels is the interaction between sensor view angle and plant canopy geometry. If, for example, an agricultural row crop is viewed from nadir or along the direction of the rows, a mixture of soil and crop reflectance would be seen (assuming the IFOV of the sensor includes at least a single row and furrow), whereas if it is viewed perpendicular to the rows at an off-nadir angle, primarily crop reflectance would be seen. Obviously, this effect also depends on the height of the crop and the distance between rows. Geometric models that incorporate these parameters have been applied to the analysis of this problem by Jackson et al (1979). A system such as the Landsat MSS with a maximum off-nadir angle of ±5.8° is obviously less affected by this problem than would be an aircraft sensor with a larger FOV.

Within-class reflectance variability

Some variation in spectral characteristics is inevitable, even for well-defined and homogeneous classes. Experimental data acquired by Duggin (1974) showed a 7 to 18 percent variability in spectral reflectance for a relatively inert material such as sandstone. Spectral variability of vegetation, caused

by such factors as plant health, age, water content and soil
mineral content, can be much larger. Even under relatively
well-controlled laboratory measurement conditions, the reflectance of corn leaves has been found to vary as much as ±17
percent at a wavelength of 0.67 μm (Landgrebe, 1978).

3.3 Classification Training

The first step of any classification procedure is the
training of the computer program to recognize the class signatures of interest. This aspect of classification is critical
for the success of the entire process and often occupies a
majority of the analyst's time. To train the computer program,
we must supply a sample of pixels from which class signatures,
e.g., mean vectors and covariance matrices (Sec. 1.5.3), can be
developed. There are basically two ways to develop signatures.
For *supervised* training, the analyst uses prior knowledge
derived from field surveys, photointerpretation, and other
sources, about small regions of the image to be classified to
identify those pixels that belong to the classes of interest.
The feature signatures of these *analyst-identified* pixels are
then calculated and used to recognize pixels with similar signatures throughout the image. For *unsupervised* training, the
analyst employs a computer algorithm that locates naturally-
occurring concentrations of feature vectors from a heterogeneous
sample of pixels. These *computer-specified* clusters are then
assumed to represent feature classes in the image and are used
to calculate class signatures. The computer-derived classes
remain to be identified, however, and they may or may not correspond to classes of interest to the analyst.

Supervised and unsupervised training thus complement each
other; the former imposes the analyst's knowledge of the area on
the analysis to constrain the results, and the latter determines

the inherent structure of the data, unconstrained by external knowledge about the area. A combination of the two techniques is often used to take advantage of the characteristics of each.

3.3.1 Supervised

For supervised training, a representative area for each desired class must be located in the image. It is important that the training area be a homogeneous sample of the respective class, but at the same time, the range of variability for the class must be included. Thus more than one training area per class is often used (Fig. 3-9a). Field surveys, aerial photographs and existing maps are used to verify the training sites. If there is considerable within-class variability, the selection of training sites can be laborious, and it is impossible to be entirely certain that a comprehensive set of training samples for each class has been specified.

In many cases it is impossible to obtain homogeneous sites. A common problem is sparse vegetation, which complicates attempts to map both vegetation and soils. One technique for improving training data under these conditions is to "clean" the sites of outlying pixels (in feature space) before developing the final class signatures (Maxwell, 1976). The cleaning operation involves applying a *threshold* operation on the training data (Sec. 3.6.1). If the cleaned training data still include more than one prominent distribution, typically evidenced by a multimodal class histogram, the common mathematical assumption of normal distributions will be violated and the classification accuracy is likely to be reduced.

One important statistical aspect of selecting training data is that a sufficient number of pixels must be used to estimate the class signature properties accurately. If a Bayes maximum-likelihood classifier is used and normal class distributions are assumed, the class mean vectors and covariance matrices must be

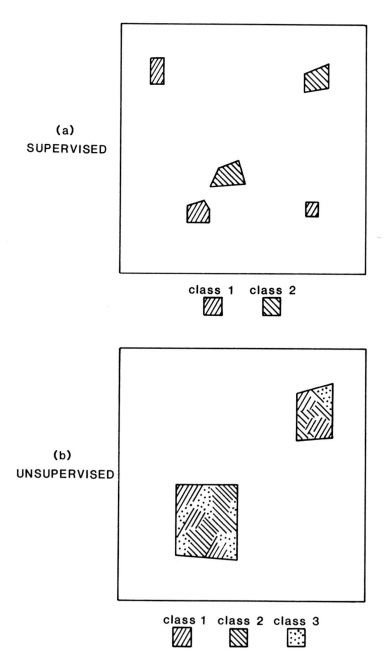

FIGURE 3-9.   Example sites for supervised and unsupervised
              training.

calculated. If K features are used, the training set for each class must contain at least K + 1 pixels to calculate the inverse of the covariance matrix [Eq. (1-38)]. To obtain *reliable* class statistics, however, 10 to 100 training pixels per class *per feature* are typically needed (Swain, 1978). The number of training pixels required for a given signature accuracy increases with an increase in the within-class variability.

3.3.2 Unsupervised

In defining image areas for unsupervised training, the analyst does not need to be concerned with the homogeneity of the sites. Often, the sites are purposely chosen to be heterogeneous to insure that all possible classes and their respective within-class variabilities are included (Fig. 3-9b). The pixels within the training areas are submitted to a *clustering* algorithm that determines the "natural" groupings of the data in the K-dimensional feature space. Each cluster then is assumed to represent the probability distribution for one class. The assignment of identifying labels to each class may be done by the analyst at this point or after classification of the full image. Because unsupervised training does not necessarily require any information about the area being classified, beyond what is in the image itself, it may be useful for delineating homogeneous areas for potential supervised training sites.

The determination of intrinsic clusters in the training data can be made in numerous ways. One of the more common methods is the K-means algorithm, also known as the Isodata algorithm (Duda and Hart, 1973). Figure 3-10 illustrates application of the K-means algorithm to a two-dimensional set of test data. These data are the same as those pictured in Fig. 1-23 and consist of three normal distributions with different mean vectors and covariance matrices. Because the data are simulated in the probability density domain as *exact* normal distributions,

there is no finite sample size error in estimating the probability density functions in this example.

In the first step of the algorithm, an initial mean vector
("seed") is arbitrarily specified for each of K classes.   Each

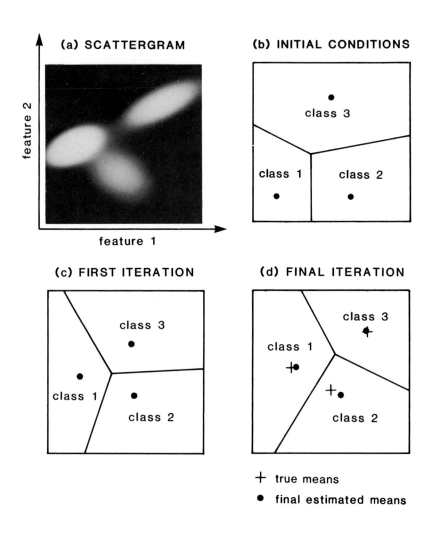

FIGURE 3-10. Clustering by the K-means algorithm.

pixel of the training set is then assigned to the class whose mean vector is closest to the pixel vector (the minimum–distance classifier) forming the first set of decision boundaries (Fig. 3-10b). A new set of class mean vectors is then calculated *from the results of the previous classification* and the pixels are reassigned to the classes (Fig. 3-10c). The procedure continues until there is no significant change in pixel assignments from one iteration to the next. In this example, the algorithm does not converge to the exact class means (Fig. 3-10d) because the estimated class means are calculated from distributions that are *truncated* by the class partitions. The algorithm is relatively insensitive to the initial choice of cluster mean vectors, but more iterations may be required for convergence if the seed vectors are not close to the final mean vectors (Fig. 3-11). The final class mean vectors may be used to classify the entire image with a minimum–distance classifier, or the covariance matrices of the clusters may be calculated and used with the mean vectors in a maximum–likelihood classification.

The number of ways to determine natural clusters of data has been limited only by the ingenuity of researchers in defining cluster criteria (such as the simple nearest–mean distance used above). Both Fukunaga (1972) and Duda and Hart (1973) describe several clustering criteria and Anderberg (1973) and Hartigan (1975) provide Fortran computer programs for many clustering algorithms. Virtually all of the commonly used algorithms use iterative calculations to find an optimum set of decision boundaries for the given data.

3.3.3 Combination

Because supervised training does not necessarily result in class signatures that are numerically separable in feature space, and because unsupervised training does not necessarily result in classes that are meaningful to the analyst, a combined

approach has the potential to meet both requirements. If time
and financial resources permit, this is undoubtedly the best
procedure to follow.

First, unsupervised training is performed on the data and
an initial classification map of the training area is produced

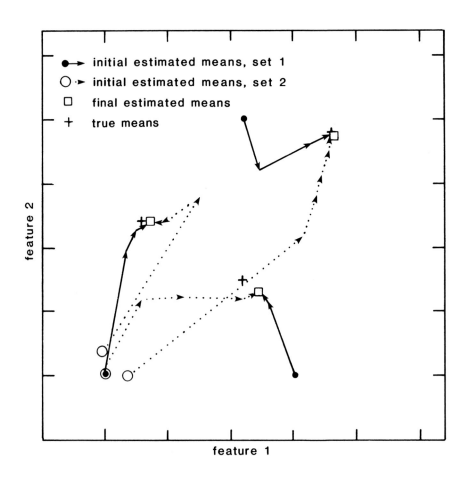

FIGURE 3-11.   Convergence of class mean estimates with the K-
               means algorithm for two different initial condi-
               tions. Each arrow indicates the estimated class
               mean at a particular iteration.

using the separable clusters defined by the computer. The analyst then evaluates the map with field survey data, aerial photographs, and other reference data and attempts to relate the feature space clusters in the training data to meaningful mapping units. Normally, some clusters must be subdivided or combined to make this correspondence; this step in the procedure is thus supervised. The revised clusters are then used as training data for the classification algorithm which is then applied to the full image. The resulting map should combine the best features of supervised and unsupervised training.

3.4 Pre-Classification Processing and Feature Extraction

Those aspects of remote sensor imagery that are used to define mapping classes are known as *features*. The simplest features, the pixel gray levels in each band of a multispectral image, are not necessarily the best features for accurate classification. They are influenced by such factors as atmospheric scattering and topographic relief, and are often highly correlated between spectral bands, resulting in the inefficient analysis of redundant data. Furthermore, more complex features derived from an image, such as measures of spatial structure, may provide more useful information for classification. Thus it is prudent to consider various pre-classification manipulations and transformations to extract the greatest amount of information from the original image. The value of pre-classification processing was recognized in the analysis of the first aerial MSS data, before the advent of Landsat (Kriegler et al, 1970; Crane, 1971; Smedes, 1971), and it continues to be important for data normalization and feature extraction.

3.4.1 Atmospheric Correction

Atmospheric scattering of visible wavelength radiation affects all satellite and high altitude imagery by reducing the modulation of the scene radiance viewed by the sensor. In the absence of an atmosphere, the modulation of the radiance, L, arriving at the sensor would be that due to the reflectance modulation on the ground

$$M_0 = \frac{L_{max} - L_{min}}{L_{max} + L_{min}}$$

$$= \frac{K\rho_{max} - K\rho_{min}}{K\rho_{max} + K\rho_{min}}$$

$$= \frac{\rho_{max} - \rho_{min}}{\rho_{max} + \rho_{min}} \tag{3-1}$$

where K is a constant given by the solar irradiance at the ground, E_G, divided by π (Slater, 1980), and ρ is the ground reflectance, all measured for the spectral band of interest. The atmosphere scatters some radiation, L_a, back to the sensor before it reaches the ground, with a corresponding reduction in the solar irradiance at the ground to E_G'. In addition, radiation reflected from the ground is attenuated by the atmospheric transmission factor, τ_a, before reaching the sensor. Thus, *in the presence of the atmosphere*, a satellite sensor sees a modulation given by

$$M' = \frac{(L'_{max} + L_a) - (L'_{min} + L_a)}{(L'_{max} + L_a) + (L'_{min} + L_a)}$$

$$= \frac{L'_{max} - L'_{min}}{L'_{max} + L'_{min} + 2L_a}$$

$$= \frac{\rho_{max} - \rho_{min}}{\rho_{max} + \rho_{min} + 2L_a/K'} \tag{3-2}$$

where $$K' = E_G' \tau_a / \pi \tag{3-3}$$

Because K' and L_a are both positive quantities, we see that

$$M' < M_0$$

i.e., the scene radiance modulation is reduced by the atmosphere. The image irradiance modulation is reduced by the same amount, and if the sensor's noise level remains the same, there is a reduction in the image SNR. This simple model describes the major atmospheric effect on satellite or high-altitude imagery, but the actual situation is considerably more complex because of scattering downward to the ground, scattering out of and into the IFOV from surrounding areas, etc.

There are other atmospheric properties that can interfere with remote sensing of the earth's surface, such as the absorption of radiation by water vapor at near IR wavelengths, a factor that can affect the band 7 image from the Landsat MSS (Pitts et al, 1974). There is also turbulence in the atmosphere that results in a random atmospheric PSF, degrading the image at higher spatial frequencies. The effect on imagery from sensors with an IFOV larger than a meter is negligible, however.

A correction for atmospheric scattering is necessary if

(1) The scattering level varies from one part of the image to another. An example is an image of a large urban area and surrounding natural areas. The image contrast and spectral characteristics of the urban area will be different from those of non-urban areas because of particulate and gaseous components in the air.

(2) A set of multitemporal images is to be analyzed and the scattering level varies with time. The changing atmospheric conditions can prevent "extension" of class signatures from one date to another.

(3) Certain types of processing are to be performed on the data, such as spectral band ratios. The radiance bias, L_a, caused by atmospheric scattering is not removed by the ratioing of spectral bands (Sec. 3.4.2).

The importance of atmospheric corrections in situations (1) and (2) has been pointed out by Fraser et al (1977), who also noted the value of *retraining* for classification under varying atmospheric conditions. Many experiments in atmospheric measurements, such as those of Dana (1978) and Hulstrom (1974), require additional ground or aircraft-based radiometric data. Because of the complexity and difficult logistics of such experiments, researchers have looked at ways of estimating atmospheric scattering levels using the image data alone.

A common atmospheric correction technique is based on the histograms of multispectral images containing deep water bodies or topographic shadows (Chavez, 1975; Potter and Mendlowitz, 1975). If the full scene gray level histograms for each band of a Landsat MSS image are plotted on the same graph, their relative positions are typically as shown in Fig. 3-12. Band 7, the near IR band, usually has some pixels with a zero, or nearly zero, gray level in water bodies or shadows. This is empirical verification that there is no measurable atmospheric scattering contribution to the image of band 7. If it is assumed that the displacement of the low end of the other histograms is due to a scattering component, then the other bands would also have some pixels with a zero gray level if it were not for the atmosphere (remember that shadows on the moon, where there is no atmosphere, are totally black). The minimum pixel gray level that can be considered statistically valid in each band, indicated by A_i in Fig. 3-12, is therefore assumed to be the atmospheric component for that band and is *subtracted from all pixels* in the image to perform the atmospheric correction.

This convenient correction is valuable in obtaining approximately corrected radiance values for spectral band ratio analysis or multitemporal normalization. As it is commonly applied,
however, a constant scattering level is assumed throughout the
scene. The technique obviously could be adapted, however, to
scattering variations over smaller regions if each region contained water or shadows. Also, if the images are to be used

(a) UNCORRECTED

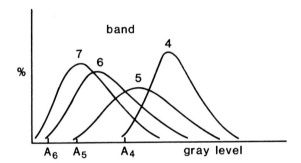

(b) CORRECTED

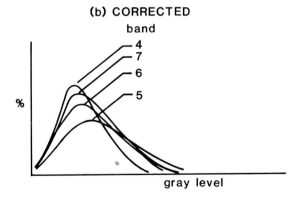

FIGURE 3-12. *Correction for atmospheric scattering using
multispectral image histograms.*

only for visual analysis of single bands or color composites, the global atmospheric correction is redundant, because the same type of bias subtraction is usually part of contrast enhancement (Section 2.2).

The correction of satellite images for atmospheric effects is still an active research area in remote sensing and probably will remain so for a long time because of the complexity of the problem. The correction technique described above, although simple, cannot be applied to images that do not have water bodies or shadows. Researchers will continue to investigate such image-based approaches, however, because they do not require auxiliary measurements on the ground or from aircraft.

3.4.2 Multispectral Ratios

One of the earliest pre-classification techniques applied to remote sensing images was the division of one spectral band by another. This type of processing can

(1) Remove temporally or spatially-varying gain and bias factors. This is accomplished only if these factors are the *same* in the bands used in the ratio.

(2) Suppress radiance variations arising from topographic slope and aspect.

(3) Enhance radiance differences between soils and vegetation.

A gain factor, a, that is the same in two bands can be removed by simply dividing each pixel in one band by the corresponding pixel in the other

$$R_{12} = GL_1/GL_2$$

$$= a\rho_1/a\rho_2$$

$$= \rho_1/\rho_2 \tag{3-4}$$

Topographic slope and aspect are common sources of this gain factor. The ratio image thus depends only on the scene reflectance ρ in the two bands, *if* there are no atmospheric effects. A common bias factor, b, can be removed by computing a ratio of the differences between three bands, taken in pairs

$$
R_{123} = \frac{GL_1 - GL_2}{GL_1 - GL_3}
$$

$$
= \frac{(a\rho_1 + b) - (a\rho_2 + b)}{(a\rho_1 + b) - (a\rho_3 + b)}
$$

$$
= \frac{\rho_1 - \rho_2}{\rho_1 - \rho_3} \tag{3-5}
$$

A common gain factor is also removed by this difference ratio. An example source of this bias factor is atmospheric scattering [see Eq. (3-2)] However, the scattering level is *not* the same in different spectral bands and Eq. (3-5) therefore would provide only an approximate correction.

An example of the amount of topographic correction that is possible with a simple spectral band ratio, even without an atmospheric correction, is shown in Fig. 3-13b. The severe shading caused by the topography in this Landsat image of the Grand Canyon, Arizona, is almost totally removed in the ratio. An image has thus been derived that is more representative of surface cover properties and, in conjuction with the ratios between other bands, would provide much better classification accuracies than would the original spectral bands. The dynamic range of the ratio image is normally much less than that of the original image because the radiance extremes caused by topography have been removed; thus, the reflectance contrast between surface cover types can be enhanced in visual displays, such as

FIGURE 3-13.   Spectral band ratios. (a) Band 5 and band 7.
              (b) R_{75}. (c) TVI_1.

color composites of different spectral ratios (Chavez et al., 1982).

Spectral ratios have been investigated extensively for measuring vegetation cover density. Maxwell (1976) and Tucker (1977, 1979), for example, found strong correlations between the ratio of Landsat MSS bands 7 and 5 and the amount of living biomass on the ground. This characteristic of the band 7-to-band 5 ratio, R_{75}, is evident in Fig. 3-13b. The North Rim of the Canyon, which is heavily forested with conifers, appears as the lightest area (highest ratio value) and the sparsely-vegetated areas within the Canyon appear dark. Lines of constant R_{75} value, i.e., isoratio contours, are shown super-imposed on a typical band 7-versus-band 5 scattergram in Fig. 3-14. This plot clearly shows how higher ratio values occur in the vegetation portion of the scattergram and that there is a continuum of ratio values down to about one. This lower bound

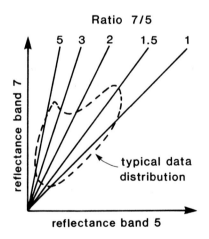

FIGURE 3-14. *Isoratio contours and a typical band 7 versus band 5 reflectance scattergram.*

on the scattergram is approximately the *soil brightness axis* defined in Sec. 3.4.4. Shadows and water will sometimes cause an extension of the scattergram below this line.

A transformed vegetation index (TVI) given by Rouse (1973) has been frequently applied to biomass estimation for rangelands and has been compared to other vegetation indices for this application by Richardson and Wiegand (1977). Two useful indices derived from Landsat MSS data are given by

$$TVI_1 = \sqrt{\frac{GL_7 - GL_5}{GL_7 + GL_5} + 0.5}$$

$$= \sqrt{\frac{R_{75} - 1}{R_{75} + 1} + 0.5}$$

and

$$TVI_2 = \sqrt{\frac{GL_6 - GL_5}{GL_6 + GL_5} + 0.5}$$

$$= \sqrt{\frac{R_{65} - 1}{R_{65} + 1} + 0.5} \tag{3-6}$$

where the 0.5 bias term automatically prevents negative values under the square root for most images. As seen from Eq. (3-6), the TVI is simply a transformation of a spectral band ratio, and consequently contains no additional information. Its advantage over the simple ratio is that in some situations the TVI tends to be approximately linearly proportional to biomass, thus simplifying regression calculations. The TVI image shown in Fig. 3-13c is visually quite similar to the simple ratio image in Fig. 3-13b. In both cases the pixel gray levels are related to the amount, type, and vigor of vegetation.

Two variations on the simple band-to-band ratio, used primarily for enhanced visual display, are the logarithmic

transformation (Goetz et al, 1975) and the arc tangent transformation (Wecksung and Breedlove, 1977).   Both the log and arc
tangent transformations of a ratio image stretch the contrast of
image areas where the ratio values are small (Fig. 3-15).   The
arc tangent transformation is similar to the TVI transformation,
although the latter was not intended for image enhancement
purposes.

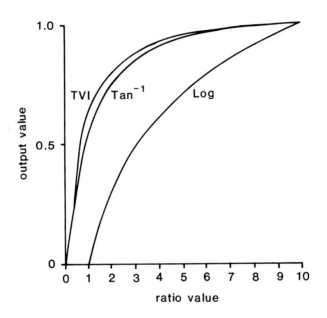

FIGURE 3-15.   Common transformations for multispectral
ratio images.

3.4.3 Principal and Canonical Components

It has been frequently observed that the individual bands
of a multispectral image are commonly highly correlated, i.e.,
they are visually and numerically similar (Fig. 3-16).   This
correlation arises from:

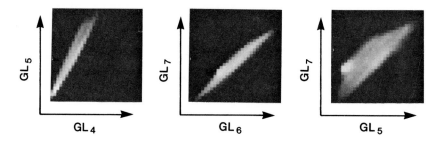

FIGURE 3-16   *Two-dimensional scattergrams between spectral bands*
for a Landsat MSS image (Schowengerdt, 1980; re-
produced with permission from Photogrammetric Eng.
and Remote Sensing, October 1980, ©1980, Am. Soc.
of Photogrammetry).

(1) Natural spectral correlation.   This correlation is
caused by, for example, the relatively low reflectance
of vegetation in Landsat MSS bands 4 and 5, and relatively high reflectance in bands 6 and 7.

(2) Topographic slope and aspect.   For all practical
purposes, topographic shading is the same in all solar
reflectance bands and can even be the dominant image
contrast component in mountainous areas and at low sun
angles.

(3) Overlap of spectral sensitivities between adjacent
spectral bands.   This factor normally is reduced as
much as possible in the MSS design and engineering
stage, but can seldom be eliminated completely.

Analysis of the original spectral bands, therefore, can be
extremely inefficient in terms of the actual amount of non-
redundant data present in the multispectral image.

   Principal and *canonical component* transformations are two
pre-classification techniques for removing or reducing this

spectral redundancy (Jenson and Waltz, 1979). They are similar in that they both form a new K-dimensional set of data from a *linear combination* of the original K features (for example, the K spectral bands of a multispectral image). The transformed features are given by

$$x_i' = \sum_{j=1}^{K} \omega_{ij} x_j \quad , \quad i,j = 1, \ldots, K \qquad (3\text{-}7)$$

where j and i denote the feature axes in the original and transformed data, respectively, and ω_{ij} are the weights applied to the original data x_j. This linear transformation may be written in vector notation as

$$X' = \mathbf{W}X \qquad (3\text{-}8)$$

where X and X' are the original and transformed K-dimensional vectors and $\mathbf{W}$ is the K-by-K transformation matrix. The principal components transformation (also known as the Karhunen-Loeve [KL] transformation) is a special case of Eqs. (3-7) and (3-8) that is optimum in the sense that the particular $\mathbf{W}$ that *diagonalizes* the covariance matrix of X is used. The principal component images, therefore, are *uncorrelated* and are ordered by decreasing gray level variance, i.e., x_1' has the largest variance and x_K' has the lowest. The result is removal of the correlation that was present between the axes of the original K-dimensional data, with a simultaneous compression of image variance into fewer dimensions.

Figure 3-17 shows that the principal components transformation in two dimensions is a rotation of the original coordinate axes to coincide with the directions of maximum and minimum variance in the data. If the mean of the data is subtracted, the origin shifts to the center of the distribution as shown.

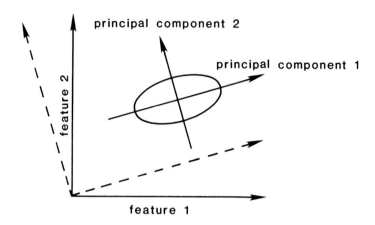

FIGURE 3-17. *Principal components transformation of a single bivariate distribution.*

If the distribution is multimodal, the nature of the transformation is the same but not as easily visualized.

An example of a principal components transformation of a Landsat MSS image is shown in Fig. 3-18. The positive-negative relationship of the first two components, a common characteristic of this transformation, expresses the uncorrelated nature of the new coordinates. The third and fourth components typically contain less image structure and more noise than the first two, indicating the data compression that has occurred. The contrast of each of the images in Fig. 3-18 has been stretched for optimum display, but their variances reveal the redistribution of image contrast achieved by the principal components transformation (Table 3-1).

FIGURE 3-18. *Principal components transformation of a Landsat MSS image. (a) Band 4. (b) PC 1. (c) Band 5. (d) PC 2. (e) Band 6. (f) PC 3. (g) Band 7. (h) PC 4.*

TABLE 3-1

Gray Level Variances for the Images in Fig. 3-18

Spectral band	Variance	% Total	Principal component	Variance	% Total
4	74.2	12.6	1	553.3	94.1
5	249.9	42.5	2	29.9	5.1
6	219.5	37.3	3	3.7	0.6
7	44.5	7.6	4	1.2	0.2

The concentration of image information in the first two principal components is typical of Landsat MSS data and implies that the *intrinsic dimensionality* of Landsat MSS imagery is about two. Figure 3-19, from Landgrebe (1978), depicts the classification accuracy of a set of 12 channel multispectral data classified in principal components form, as a function of the number of components used in the classification. Note that by using the first 3 principal components, a factor of 4 can be saved in classification computation time with little loss in accuracy. The computer time required to calculate the principal components must be considered in a complete comparison, however.

Whereas the principal components transformation does not utilize any information about class signatures, the canonical transformation maximizes the *separability* of defined classes. The class means and covariance matrices must be specified for the transformation; the average *within-class* covariance matrix is calculated from the individual class covariance matrices and the *between-class* covariance matrix is calculated from the class mean vectors [Eq. (1-34) with appropriate substitutions]. A

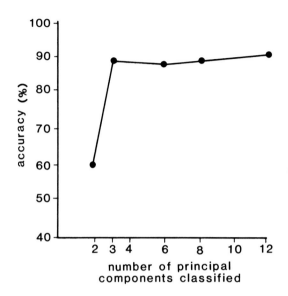

FIGURE 3-19. *Classification accuracy for a set of principal components imagery (Landgrebe, 1978; from Remote Sensing: The Quantitative Appraoch, Swain and Davis, eds., ©1978, McGraw-Hill Book Co. Used with the permission of McGraw-Hill Book Co.)*

transformation matrix, **W** , is then found that simultaneously diagonalizes the between-class covariance matrix and transforms the average within-class covariance matrix to the identity matrix, i.e., a diagonal matrix with all variances equal to one. The desired goal is to maximize the separability between classes and minimize the variance within classes. This result is only approximately achieved in practice, however, because the within-class covariance matrices for different classes are *not* equal to each other for real data. Thus, the variance can only be minimized for the *average* within-class covariance matrix.

Figure 3-20 depicts the canonical transformation for two classes with equal covariance matrices in two dimensions. For two classes, the first canonical axis must pass through the class means (this is not true for more classes, of course) and, for the example shown, the two classes can be separated easily by using only the first canonical axis. Both within-class covariance matrices have been diagonalized, i.e., ρ_i equals zero (Fig. 1-21) and the variances have been equalized because the two classes originally had equal covariance matrices.

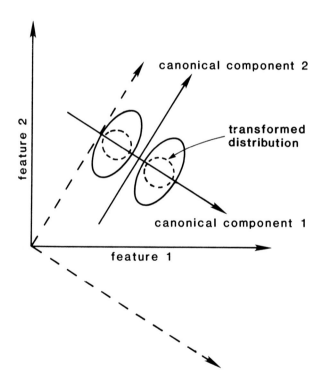

FIGURE 3-20. *Canonical components transformation of two bivariate distributions with equal covariance matrices.*

In a rangeland vegetation mapping application, Maxwell (1976) has shown that a classification using only the first canonical feature resulted in less error than one based on the best discriminating feature, R_{75}, among the four Landsat bands and two ratios, R_{75} and R_{54}. Canonical transformations also have been useful for visual geologic mapping from color composite displays of the canonical feature images (Blodget et al, 1978).

3.4.4 Vegetation Components

A third type of linear feature transformation, designed specifically for agricultural biomass monitoring, was first proposed by Kauth and Thomas (1976). They noted that the gray level scattergrams between pairs of Landsat bands for numerous agricultural scenes exhibit certain consistent properties, for example, a triangular-shaped distribution between band 7 and band 5 (Fig. 3-16). Visualization of these distributions in K dimensions (K equals four for the Landsat MSS) gives a shape described by Kauth and Thomas as a "tasseled cap" whose base they called the "plane of soils."

Using a *particular set of data*, Kauth and Thomas first derived a linear transformation, Eq. (3-8), of the four Landsat MSS bands that would yield a new axis called "soil brightness index" (SBI) defined by the signature of non-vegetated areas. A second coordinate axis, orthogonal to the first and called "greeness vegetation index" (GVI), was then derived to point along the direction of vegetation signatures. A third and fourth transformed axes, called "yellow stuff" and "non-such", respectively, also were derived to be orthogonal to the first two axes. The transformation coefficients for their original set of data from Landsat-1 are given in Table 3-2, along with coefficients that were later derived for Landsat-2. It is

important to note that these transformed axes are orthogonal
only in the four transformed dimensions and are not orthogonal
in a two-band space, as shown in Fig. 3-21. A perpendicular
vegetation index (PVI) that *is* orthogonal to the soil line in
two dimensions has been defined by Richardson and Wiegand (1977)
and a generalization of vegetation indices in K dimensions has
been described by Jackson (1983).

The purpose of these transformations is to obtain a vegetation indicator that is independent of soil background effects
and can be used to monitor the production of biomass in agriculture (Thompson and Wehmanen, 1979). For example, a plot of the
GVI as a function of time for two different crops shows two
similar bell-shaped curves, separated in time by the difference
in planting and growing cycles (Fig. 3-22). It is possible to
numerically model these curves with relatively simple parametric

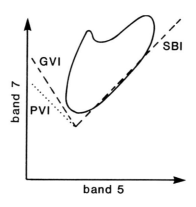

FIGURE 3-21. *Projection of vegetation components onto two
dimensions.*

TABLE 3-2

Tasselled Cap Transformation Coefficients ω_{ij}[1]

Landsat-1			
MSS Band			
4	5	6	7
soil brightness 0.433	0.632	0.586	0.264
greeness −0.290	−0.562	0.600	0.491
yellow stuff −0.829	0.522	−0.039	0.194
non-such 0.223	0.120	−0.543	0.810

Landsat-2			
MSS Band			
4	5	6	7
soil brightness 0.332	0.603	0.676	0.263
greeness −0.283	−0.660	0.577	0.388
yellow stuff −0.8995	0.428	0.0759	−0.0408
non-such −0.0159	0.428	−0.452	0.882

[1]These coefficients are for a 0–63 scale in band 7 and a 0–127 scale in the other bands, i.e., the same scales used for the data on computer compatible tapes (CCTs).

functions, obtain the parameters (for example, width, height and temporal offset) that describe a particular temporal curve and use these parameters as features in a classification for crop types (Badhwar, 1982). The transformed coordinates of yellow stuff and non-such have been shown to indicate changes in atmospheric haze conditions and, therefore, may be useful for relative calibration of images for atmospheric differences (Malila et al , 1980).

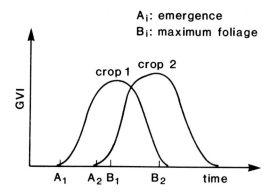

FIGURE 3-22.   *Temporal behavior of the greeness component for two different crops.*

The transformation derived by Kauth and Thomas for agriculture has been applied widely to many other types of vegetation. Although the transformed variables of soil brightness and vegetation greeness retain their general meaning in these applications, the transformation coefficients should be rederived for conditions that differ greatly from those of the original analysis of Kauth and Thomas.

3.4.5 Spatial Information

Historically, photointerpreters have had preeminence in the use of spatial information for remote sensing mapping. It is only relatively recently that there has been sufficient quantitative understanding of spatial image structure to permit its use in computer classification. Analysis of spatial information necessarily requires analysis of pixels within at least a local neighborhood. Therefore, just as spatial filtering is more

computation-intensive than contrast enhancement, spatial information extraction algorithms generally require considerably more
computer time than do single pixel algorithms, such as spectral
band ratios.

One quantitative description of spatial information that
has received considerable attention in remote sensing is
texture. There is no definitive characterization of texture,
but all numerical definitions are related to the "coarseness"
and contrast of image detail. The concept of spatial texture is
illustrated by the artificial patterns and aerial photograph in
Fig. 3-23. In the aerial photograph, different textural characteristics are exhibited by different land uses, from urban residential areas (left center) to agriculture (upper right).

Texture is usually defined as some local property of an
image, i.e. a measure of the relationships between the pixels in
a neighborhood. At first it would seem that simple statistical
quantities, such as the local gray level variance, would be
valid measures of texture. They are, to the extent that they
express local image contrast, but they do not incorporate information about the spatial frequency characteristics of the image.
The essential difference between the residential and agricultural areas in Fig. 3-23b, for example, is that the former is
characterized by relatively high frequency detail. Thus, any
valid texture feature must somehow include both contrast and
frequency information.

One approach to texture feature extraction is based on the
gray level spatial-dependence matrix (sometimes called the
cooccurrence or transition matrix; Haralick et al , 1973). This
matrix denotes the probabilities of transition from one gray
level to another between neighboring pixels in the image. A
window algorithm is used (Fig. 1-15) and for each position of
the window, the number of times a pixel with gray level GL_i
occurs next to one with gray level GL_j is determined and used as

(a) SYNTHETIC PATTERNS (Pratt, 1978)

(b) AERIAL IMAGE PATTERNS

FIGURE 3-23. Examples of spatial texture.

the (i,j) element in the spatial-dependence matrix. Areas of low contrast or low spatial frequency will thus produce a concentration of counts near the diagonal of the spatial-dependence matrix, i.e. where i is close to j. Areas of high contrast and high spatial frequency redistribute this concentration away from the diagonal. A total of 13 scalar texture features derived from the spatial-dependence matrix were defined by Haralick et al (1983). These features are generally measures of the location and degree of concentration of pixel counts in the matrix.

In addition to spectral features, each pixel may also have one or more texture features (as many as 13 in the above algorithm) for each spectral band. If texture directionality is important, this dimensionality can increase at least two-fold and data compression techniques, such as principal components (Sec. 3.4.3), are commonly applied to reduce this high dimensionality. Both Haralick et al (1973) and Wiersma and Landgrebe (1976) reported improvements in average classification accuracy of about 10 percent for Landsat MSS data after the spectral features were augmented with a subset of the spatial-dependence matrix features.

Spatial information in images can be quantitatively expressed in other ways. For example, spatially homogeneous image areas can be found by using simple image processing techniques such as thresholding and smoothing of gradient images (Sec. 2.3.3) as shown by Haralick and Dinstein (1975). Similarly, the ECHO (Extraction and Classification of Homogeneous Objects; Kettig and Landgrebe, 1976), AMOEBA (Bryant, 1979), and BLOB (Kauth et al, 1977) algorithms segment an image into regions on the basis of spatial and spectral homogeneity, i.e., *lack* of texture. In the ECHO algorithm, for example, adjoining pixels are initially aggregated into small cells (e.g. 2-by-2 pixels) based on the similarity of their spectral vectors. Cells that cross spatial boundaries (edges) are detected by a threshold on the cell variance in each spectral band, and the pixels in the cell are not aggregated if an edge is detected within the cell. The homogeneous cells found at this stage are then aggregated further if they are spectrally similar to neighboring cells. The resulting spatially homogeneous areas are then classified in their entirety as *single* spectral samples, rather than pixel-by-pixel. Kettig and Landgrebe (1976) evaluated several variations on the basic algorithm and obtained a substantial increase in accuracy over a simple pixel-by-pixel classifier for an agricultural area. This type of technique would be expected to work best where the scene consists of relatively homogeneous objects that are several pixels large (Fig. 3-24).

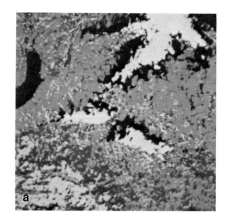 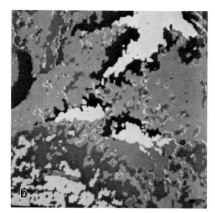

FIGURE 3-24. Example application of the ECHO algorithm. (a) Pixel-by-pixel classification. (b) ECHO classification. (Lindenlaub and Davis, 1978; from Remote Sensing: The Quantitative Approach, Swain and Davis, eds., ©1978, McGraw-Hill Book Co. Used with the permission of McGraw-Hill Book Co.)

Other techniques for incorporating spatial information attempt to define the *context* of objects within a scene, and often utilize a *syntactic* image description, i.e., an image "language." Syntactic classification is equivalent to making statements of the type, "If this object belonging to this class occurs here in the image and another, unknown object occurs in this location relative to the first object, then the second object must belong to this class." Although the logic of this type of image analysis is very appealing, the algorithms required to implement it are complex. Some experiments in remote sensing using syntactic image description have been performed by Brayer et al (1977) and others. Wharton (1982) described a relatively simple context algorithm that uses the local proportions of different surface cover classes, obtained from a pixel-by-pixel multispectral classification, to determine a specific land use class for a neighborhood.

In any attempt to extract spatial information from images it should be remembered that spatial image structure in remote sensing is caused by both topographic shading arising from terrain slope and aspect and by surface cover variations in reflectance. In many cases the topographically-induced spatial variations are not of interest and in fact represent a source of noise in the analysis. Because spectral ratio processing (Sec. 3.4.2) suppresses topographic shading, it can be a valuable preprocessing step before application of spatial analysis algorithms to extract spatial information about reflectance variations.

3.5 Classification Algorithms

Classification algorithms may be grouped into one of two types, *parametric* or *nonparametric*. Parametric algorithms assume a particular class statistical distribution, commonly the normal distribution, and then estimate the parameters of that distribution, such as the mean vector and covariance matrix, to use in the classification algorithm. Nonparametric algorithms, on the other hand, make no assumptions about the class distributions. Nonparametric techniques are sometimes termed *robust* because they work well for a wide variety of class distributions, *if* the class signatures are reasonably distinct to begin with. Of course, parametric algorithms usually yield good results under the same conditions, even if the assumed class distribution is invalid.

The straightforward implementation of the maximum-likelihood classifier, a parametric approach, requires calculation of the probability that each pixel belongs to each of the defined classes. Each pixel is then assigned to that class for which the probability is the greatest. It is intuitively obvious, however, that the probability of a pixel belonging to some classes is extremely small (for example, a "water pixel" will have a significantly different signature than a "vegetation pixel" in Landsat MSS data) and the actual classification decision, therefore, usually will be made between relatively high probabilities for two or three classes. Furthermore, any classification actually consists of a one-time calculation of decision boundaries, followed by a comparison of each pixel's feature vector and the location of those boundaries. By incorporating these facts in classification algorithms, their efficiency can be improved by an order of magnitude or more, with little or no reduction in accuracy.

3.5.1 Efficient Algorithms

Figure 3-25 depicts a hypothetical scattergram for three classes. All the pixels belonging to class 3 and most of the pixels in classes 1 and 2 can be classified easily by comparison of the location of each pixel's feature vector relative to class boundaries that are parallel to the feature space axes; hence,

this type of classifier is called the *level slice* algorithm
(Fig. 3-25a).   It is a K-dimensional extension of gray level
thresholding (Sec. 2.2.3) and can be easily implemented in
digital hardware for interactive displays.   The more general
situation where the opposite sides of the class boundaries are
parallel to each other, but not necessarily to the feature axes,
results in the *parallelepiped* algorithm (Fig. 3-25b).   Both of
these algorithms are nonparametric because no assumptions are
made about the actual probability distributions within the class
boundaries.   Classification of the pixels in the shaded area is
ambiguous with either the level slice or parallelepiped algorithms.   Addington (1975) suggested a hybrid parametric and
nonparametric approach that classifies these ambiguous pixels
with the maximum-likelihood algorithm.   For one set of Landsat
test data, Addington noted computation time was reduced by more
than a factor of ten, compared to a maximum-likelihood classification of all pixels, with only a two or three percent increase
in error for each class.   The parallelepiped classifier used
alone reduced the computation time by another factor of two, but
was not able to resolve the ambiguity in the overlapping region
between classes 1 and 2 and therefore resulted in higher classification error.

A more general approach to improving classification efficiency is the *table look-up* algorithm developed by Eppler
(1974).   This technique can be used with any classification rule
because it relies only on a description of the class partitions
and threshold boundaries.   After these are determined from the
training statistics using the desired classification rule, they
are coded in memory as functions of each feature dimension (see
Appendix C).   Once these tables are specified the algorithm
performs a series of fast inequality checks to classify a pixel.

In two dimensions the decision boundaries are stored as
four tables that describe the lower and upper decision bounda-

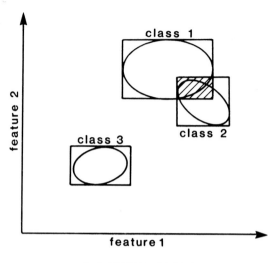

(a) LEVEL SLICE

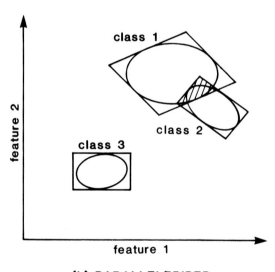

(b) PARALLELEPIPED

FIGURE 3-25.   Level slice and parallelepiped class boundaries in two dimensions.

ries for each class as a function of each of the features (Fig. 3-26). A pixel with a feature vector $(\hat{x}_1, \hat{x}_2)$ is first compared with the boundaries for class 1 in feature 1, given by $L_1(1)$ and $H_1(1)$. If $\hat{x}_1$ falls within that range, $\hat{x}_2$ is compared with the boundaries of class 1, $L_2(1, \hat{x}_1)$ and $H_2(1, \hat{x}_1)$, in feature 2. If $\hat{x}_2$ falls within that range the pixel is classified into class 1. If either the range check on $\hat{x}_1$ or $\hat{x}_2$ fails for class 1, the program proceeds to class 2. If the range checks fail for all classes the pixel is assigned to a threshold class. The algorithm can be further optimized by starting with the class of the preceding pixel, thus making use of the spatial correlation in the image, and proceeding through the classes in order of

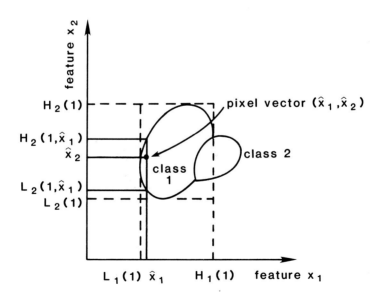

FIGURE 3-26. Table look-up algorithm in two dimensions. (after Fig. 1, Eppler, 1974).

decreasing *a priori* probabilities, if they are not assumed to be equal. The correct class at each pixel will then likely be among the first checked.

The table look-up implementation of any classifier, including the maximum-likelihood algorithm, incurs no additional error beyond that of the classifier itself. A significant amount of memory is required to store the boundary tables, however. For example, in four dimensions two additional pairs of tables are required, $[L_3,H_3]$ and $[L_4,H_4]$, that are three and four-dimensional respectively. In addition, the tables must be recalculated if there is any change in the training data or thresholds.

3.5.2 Decision Tree Algorithms

In some situations it may be easy to separate certain classes or groups of classes with the information from only one or two features. For example, water nearly always exhibits a very low radiance in Landsat MSS band 7, a fact that permits an easy separation from relatively high reflectance vegetation classes by simple gray level thresholds (Fig. 2-7). The *decision tree* (also called stratified or layered) classifier is designed to take advantage of such situations to improve efficiency and, if possible, accuracy. The decision tree classifier progresses through a series of stages, or layers; at each layer certain classes are separated in the simplest manner possible. It is a very flexible approach that permits different features *and* classification rules to be used for separating different classes. Decision tree classifiers are particularly useful for multitemporal and multisource data because of this flexibility.

Figure 3-27 shows the spectral radiance ranges for eight classes in a 13 band image from the Skylab multispectral scanner (Swain and Hauska, 1977). This data will be used to illustrate the design of a decision tree classifier. It can be seen that

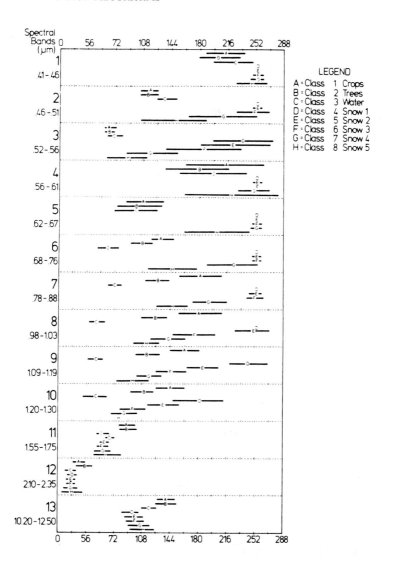

FIGURE 3-27. Spectral radiance data from the Skylab multi-spectral scanner (Swain and Hauska, 1977; ©1977 IEEE).

classes A–C can be separated from classes D–H with only the data in band 5. This operation will therefore be the first layer in the classification. Classes A–C can be separated using bands 7 and/or 8; classes D–F can be separated from G and H with only the data in band 7. These operations will therefore form the second layer of the classification. The separation of the remaining classes using different spectral bands is determined in a similar fashion until all the classes have been isolated. Figure 3–28 shows the final decision tree with the classes identified and features used at each level. In this particular example, experimentation indicated that some additional features were desirable at certain stages in the decision tree for greater accuracy; these are indicated in parentheses in Fig. 3–28.

In this example no more than three features were necessary at any stage and many of the classification steps used only one or two features. The decision tree approach was therefore much

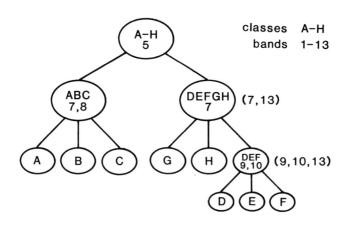

FIGURE 3-28.   Manually designed decision tree for the data in Fig. 3-27 (after Fig. 1, Swain and Hauska, 1977; ©1977 IEEE).

more efficient than a "dumb" algorithm, which would have required all six features for each class. Efficiency may be improved
further by using simple classification algorithms at some
stages. For example, the initial band 5 classification might be
accurately performed by a manually-selected gray level threshold
because the two groups of classes are well separated. There has
been some research in the automated design of decision trees,
using approaches that attempt to optimize a compromise function
of efficiency and accuracy. The drawbacks to decision tree
classifiers are that considerable analyst effort may be required
to design an effective tree and, because the decision tree is
data-dependent, a new tree must be designed for each new
application.

3.6 Post-Classification Processing and Analysis

A digital classification map usually is stored in a digital
image format with pixel gray levels of one, two, and three, for
example, representing class 1, class 2, and class 3, respectively. Although this numerical representation is commonly used,
many of the arithmetic operations described in Chapter 2 cannot
be applied to classification maps. The reason is that the gray
levels in classification maps do not represent the *magnitude* of
a quantity, but are rather a convenient coding of the pixel
labels generated by the classification process. As an example,
suppose we want to smooth a noisy classification map that has
many single pixel misclassifications (this is characteristic of
a simple pixel-by-pixel multispectral classification). A moving
average, low-pass filter as described in Section 2.3.1 will not
work because, for example, the average of class 2 and class 4 is
not class 3! The point is that algorithms for processing digital classification maps must use logical, rather than algebraic,
operations.

3.6.1 Thresholding

The specification of interclass partitions does not address the problem of "outlying" pixels, i.e., pixels that have a low *a posteriori* probability of belonging to any of the training classes. An example of how outliers can be misclassified was briefly discussed in Sec. 3.2.1. For some nonparametric classifiers, such as the parallelepiped algorithm, threshold boundaries are inherent in the definition of the classifier (Fig. 3-25). For a parametric classifier, such as the maximum-likelihood algorithm with a normal distribution assumption, outliers can be eliminated from the classification by specification of a *probability threshold*. Figure 3-29 shows probability threshold boundaries in one and two dimensions for a normal distribution.

Probability thresholds are not implemented directly on the classified pixels but on the associated *a posteriori* probability for the class of each pixel. As discussed in the introduction to this section, a logical decision to "accept" or "reject" the class at each pixel is then made based on the numerical threshold on probability. Thresholds do not improve the classification accuracy of pixels within the class boundary; they only prevent misclassification of pixels outside the boundary. Thus, accuracy estimates from test samples can not improve with the application of thresholds. Thresholds can be extremely valuable, however, in improving estimates of the total area covered by each class within a study area.

Figure 3-30 illustrates the effect of thresholds in a maximum-likelihood classification. In the original classification, the upper end of a lake (in the center of the image), which is shallow and silty, is misclassified as a geologic class (basalt) because of spectral similarity between that part of the lake and the basalt training class. The application of thresholds results in the exclusion of these pixels from the final map.

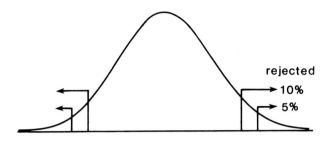

(a) ONE DIMENSION

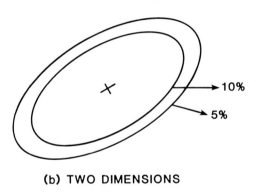

(b) TWO DIMENSIONS

FIGURE 3-29. Probability thresholds for a normal distribution.

Also note that shallow areas and land-water mixture pixels around the lake are also changed to the threshold "class" and that a few pixels, originally classified correctly as water, are similarly eliminated. Some experimentation is necessary in each particular classification to obtain effective thresholds.

(a) ORIGINAL CLASSIFICATION

(b) THRESHOLDED CLASSIFICATION

FIGURE 3-30. Example of probability thresholding.

In general, a compromise threshold that eliminates the maximum number of pixels throughout the map *without* significantly affecting the training pixels is desired.

3.6.2 Smoothing

If only spectral information is used in a pixel-by-pixel classification, the resulting maps usually have a "noisy" appearance. That is, there are many isolated pixels, or small groups of pixels whose classification is different from that of their neighbors. It is reasonable, however, to expect some degree of spatial correlation in surface cover from pixel to pixel. If this spatial information can be included in the classification process in some way, many of the isolated misclassifications can be eliminated, resulting in a *smoothed* map.

From another viewpoint, the original spectral classification may be considered accurate, i.e., not noisy, at every pixel. We may, however, not be interested in pixel-to-pixel variations. Quite often only larger scene units, specifically those that occupy areas above a certain minimum size, are important. An example is agricultural classification where we want to identify the crop type in a field and may not be interested in relatively small areas within the field.

A classification smoothing technique that uses the minimum area rationale was described by Davis and Peet (1977). The procedure alters those class-homogeneous regions that have less than the specified minimum area by examining the external boundary pixels around the region. The classification of the entire area is changed to the class that is the majority along the original boundary. A different minimum area can be specified for each class and weights can be applied to the boundary pixel classes. An example of map smoothing with this technique is shown in Fig. 3-31.

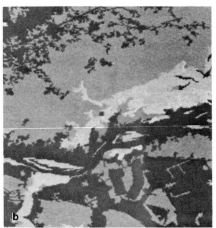

FIGURE 3-31.   Classification map smoothing with a minimum area
                constraint. (a) Original map. (b) Smoothed map
                (minimum area 25 pixels). (Davis and Peet, 1977;
                reprinted by permission of the publisher; ©1977 by
                Elsevier Science Publishing Co., Inc.)

Another classification smoothing algorithm is the *majority*
filter (Goldberg et al, 1975) illustrated in Fig. 3-32.   A
spatial window is passed through the classification map, and at
each pixel the majority class within the window is determined.
If the majority class is different than the class of the center
pixel in the window, the center pixel's classification is
changed to the majority class.   If there is no majority class in
the neighborhood, the center pixel is left unchanged.   At each
new position of the window, the original pixel classifications
are used by the algorithm, not the processed data from the
previous window position.   Although class and spatial weights
may be applied in the majority filter, it is *not* a linear
filtering operation as described in Chapter 2, and it does not
necessarily employ a minimum area constraint as in the Davis and

(a) CENTRAL PIXEL CHANGED

original classification	class	# pixels	final classification

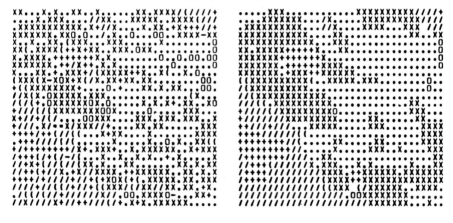

(b) CENTRAL PIXEL UNCHANGED

(c) EXAMPLE

original map   smoothed map

FIGURE 3-32. The 3-by-3 majority filter.

Peet algorithm. The results are obviously similar, however, to those in Fig. 3-31.

A direct way to obtain smoother classification maps would be to smooth the original image data before classification with a low-pass filter. This is equivalent to using a sensor with a larger IFOV. Maxwell (1976) used this approach and obtained a significant improvement in classification accuracy for a range-land area. A similar improvement has been noted in certain other applications, particularly urban land use mapping (Markham and Townshend, 1981). Two disadvantages of image smoothing are: 1) the spatial resolution of the data is clearly reduced, thus introducing more class mixture at spatial boundaries, and 2) more computation time may be required than for map smoothing because each image (feature) must be smoothed. A thorough comparison of the relative merits of this technique and the post-classification smoothing procedures has not been reported, however.

3.6.3 Error Analysis

The increasing use of quantitative mapping techniques with remote sensing images has stimulated a parallel interest in quantitative estimation of errors in the resulting maps (van Genderen et al., 1978). Although the numerical nature of computer classification facilitates error analysis, the specification of proper statistical procedures for testing map accuracy remains the most important aspect of any test procedure. A brief introduction to classification error analysis is provided here to indicate the importance of using correct statistical procedures and to point out some of the common pitfalls. There are a great many textbooks on statistics and error analysis; two particularly useful ones are the books by Davis (1973), with many Fortran programs, and Bevington (1969), a basic, but practical guide.

A complete accuracy test of a classification map would require verification of the class of every pixel. Obviously this is impossible and indeed defeats the purpose of image classification. Therefore, representative *test areas* must be used to estimate the map accuracy with as little error as possible. The map accuracy for each class is nearly always estimated by dividing the number of correctly classified test pixels in the class by the total number of test pixels in the class. The correct ("true") classes of the test pixels are determined from independent information, such as ground survey maps and aerial photography. Test areas are generally one of three types:

    (1) Training sites from a supervised classification

    (2) Analyst-specified homogeneous test sites

    (3) Randomly located pixels or sites

The use of training data for accuracy estimation is strongly biased and invariably yields overly optimistic accuracy figures. Nevertheless, accuracies calculated from training data are sometimes quoted in the literature with the implication that they indicate the overall accuracy of the classification. The only valid information supplied by the accuracies of classified training sites is an indication of the homogeneity of the training sites and the separability of the training class signatures. This information is useful for refining supervised training sites, but is not a statistically valid test of map accuracy.

The second approach to accuracy estimation, the use of analyst-specified test sites, is vastly preferable to the use of training data. During the selection of training sites, the analyst can often purposely locate more sites than are needed for training. Half may then be randomly selected and used for training and the other half reserved for testing. Because all sites are selected by the same analyst using the same information (ground surveys, aerial photography, existing maps), there is some bias in the accuracy estimates. However, they do provide an indication of the consistency of the classification if the test sites are widely dispersed throughout the study area. If there are any serious errors in the classification they almost surely will be apparent from these test data.

The final approach, the use of random pixel samples, is the least biased of the three. There are different ways to randomly select test pixels or sites; the best approach in a particular application depends on the nature of the study area and the classes being mapped. For example, the optimum sampling design for testing a classification of spatially regular agricultural fields (Bauer et al, 1977, describe a detailed procedure for agricultural classification testing) might be entirely different from that for testing a wildlands classification (Todd et al, 1980). In all random sampling procedures, however, it is desirable to select random *groups* of pixels rather than single pixels because of the practical difficulty in accurately locating single pixels on the ground or in aerial photographs, and to reduce the amount of verification work required. Thus, the test procedure commonly involves overlaying a grid on the classification map, followed by random selection of test cells, each containing at least several pixels, from the grid. The test cells are then located on the ground or in aerial photographs and the percentage coverage of each class within the cells is measured and compared to the corresponding percentages from the classification map to estimate its accuracy.

The results of classification testing are usually cross-tabulated in the form of a *contingency table* (sometimes called a confusion matrix) such as in Fig. 3-33. The values along the diagonal represent the percentage of correctly classified pixels for each class; values along a given row indicate how misclassified pixels are distributed among the classes. The sum of the values in each row is 100 percent to account for all the test

		percent test pixels			# test pixels
		map class			
		1	2	3	
	1	84.3	4.9	10.8	102
true class	2	8.5	80.3	11.2	152
	3	6.1	4.1	89.8	49

average accuracy = (84.3 + 80.3 + 89.8)/3 = 84.8%

overall accuracy = [84.3(102) + 80.3(152) + 89.8(49)] /303 = 83.2%

FIGURE 3-33. Example contingency table.

pixels. If the map has been thresholded, a fourth column is necessary to include the threshold "class." The average accuracy is the average of the accuracies for each class, and the overall accuracy is a similar average with the accuracy of each class weighted by the proportion of test samples for that class in the total test set. Thus, the more accurate estimates of accuracy, i.e., those from larger test samples, are weighted more heavily in the overall accuracy.

Confidence limits can be placed on the estimates of accuracy derived from testing. Because the classification of a pixel (in the absence of thresholds) is either right or wrong, the correctness of the class at each pixel constitutes a *binomial* population. The estimated mean probability of correct classification for each class is given by the test accuracy, as determined above, and is a random variable with a binomial probability distribution. The associated statistical error in the accuracy estimate can be found from tabulated confidence levels on estimates of the mean of a binomial distribution. A table for 95 percent confidence is given by Hord and Brooner (1976) and discussed further by Hay (1979). Figure 3-34 is a graph of data from this table that clearly shows the dependence of the confidence range on the number of test samples and the asymmetric distribution of confidence about the estimated mean.

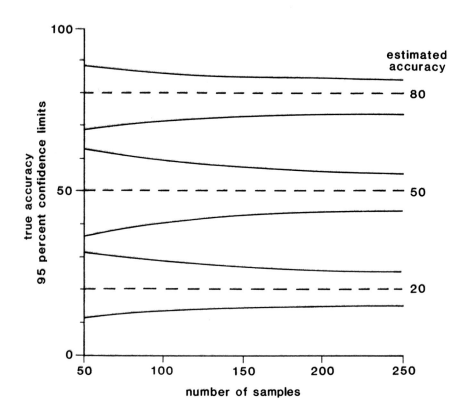

FIGURE 3-34.   Confidence bounds on estimates of the mean of a binomial distribution for various sample sizes.

From this graph we can obtain confidence ranges for the data in Fig. 3-33 (Table 3-3).

TABLE 3-3

Confidence Ranges for the Data in Fig. 3-33

Class	# Test Pixels	% Correct	95% Confidence Levels
1	102	84.3	76-90
2	152	80.3	73-85
3	49	89.8	79-95

Although class 3 has the greatest estimated accuracy, it also has the greatest uncertainty because of the small sample size. It can be seen from Fig. 3-34 that over 250 test samples are necessary to estimate the mean accuracy to within ±5 percent.

Finally, it should be remembered that the accuracy estimates obtained from classification testing are only as good as the data used to identify the classes in the test sites. The analyst should always allow for, and estimate if possible, errors in the external reference data, whether it is from field surveys, aerial photointerpretation, or existing maps. In practice, however, estimation of these errors is usually difficult, and is consequently not included in the analysis.

3.7 Non-Image Features

Most applications of classification in remote sensing use either multispectral images directly or features derived from the images (Sec. 3.4). There are, however, other valuable sources of information available in remote sensing that can be used in conjunction with image classification. One of these sources is the time domain; the relatively frequent repetitive

coverage provided by satellite sensors is invaluable for monitoring changes on the earth's surface. In fact, the temporal
information from repetitive coverage is essential for classification of crop types at the relatively low spatial and spectral
resolution of the Landsat MSS. Another useful source of information in remote sensing is digital elevation data. Elevation,
slope and aspect of the terrain often determine the type and
density of vegetation in arid and mountainous regions. If these
associations are sufficiently well-known *and* can be modeled
accurately in classification, the terrain data can be as important as imagery in mapping vegetation.

The use of these two types of non-image data in conjunction
with images in classification requires accurate registration
between two images acquired at different times (Billingsley,
1982) or between an image and a (digitized) map. The techniques
described in Sections 2.5 and 2.6 may be used directly in the
former case, but the registration of an image to digital elevation data can benefit from special preprocessing of the elevation data, as described below.

3.7.1 Multitemporal Images

The ground scene that is recorded almost instantaneously in
a single multispectral image is a *dynamic* situation in many
instances. For example, the growth and harvesting of agricultural crops are temporal phenomena, characterized by pronounced
changes in the spectral signatures of individual fields.
Natural vegetation exhibits similar, but usually less dramatic,
changes (Fig. 3-35). Other applications of remote sensing, such
as monitoring urban growth and land use change, have temporal
characteristics that extend over much longer periods of time
than vegetation cycles.

One obvious way to quantify temporal changes between two
multispectral images is to classify each image independently

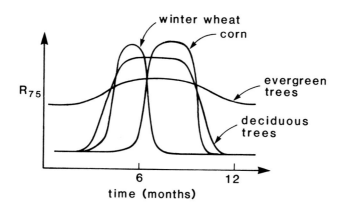

FIGURE 3-35. *Generalized temporal changes in spectral sig-*
natures for different types of vegetation in a
temperate climate.

with its own training data (Fig. 3-36a). The two classification
maps may then be compared to locate the pixels that have changed
from one class to another, thus producing a class change map.
This map may be summarized statistically by a *transition matrix*
(much like the contingency table of Fig. 3-33) that cross-
tabulates the percentage of pixels representing one class on the
first date and another class on the second date. One beneficial
aspect of this approach to multitemporal analysis is that the
independent training in each image (called *retraining* in a
temporal context) tends to normalize global differences, such as
atmospheric changes between the two images. It is an inefficient approach, however, because the amount of analyst effort
and computation time are directly proportional to the number of
image dates involved.

Another multitemporal analysis technique is "stacking" of
the images from different dates (Fig. 3-36b). A *single* set of
training signatures is then developed and the entire data set

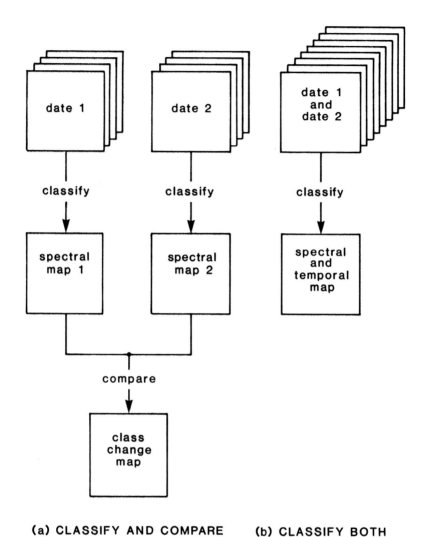

(a) CLASSIFY AND COMPARE   (b) CLASSIFY BOTH

FIGURE 3-36. *Two direct approaches to multitemporal classification.*

classified at once, as if it were a single image. Differences in atmospheric scattering and solar elevation between the images from different dates must be removed from the data set before classification. The solar elevation normalization may be achieved, after atmospheric correction, by division of the data by the cosine of the solar angle from vertical (an approximation that assumes surface cover materials are Lambertian reflectors; Slater, 1980). Two drawbacks with this multitemporal analysis procedure are that the additional images from different dates add dimensionality to the data that makes selection of supervised training sites more difficult, and the temporal and spectral features have equal status in the combined data set. Thus, spectral changes within one multispectral image cannot be easily separated from temporal changes between images in the classification map.

Swain (1978b) developed a maximum-likelihood *cascade* classifier to remove this coupling between the spectral and temporal dimensions. The cascade classifier uses the image from the first date to calculate maximum-likelihood discriminant functions that determine *a priori* probabilities for the classification of the next date. Transition probabilities between classes from one date to the next must also be estimated. Figure 3-37 shows an example of the improvement in classification accuracy obtained with the cascade classifier compared to single-date classification.

Various techniques described earlier, such as principal components (Sec. 3.4.3) and decision tree classifiers (Sec. 3.5.2), may be used to reduce or circumvent the high dimensionality of multispectral-multitemporal imagery. One approach, called *change vector analysis* (Malila, 1980) uses the Kauth-Thomas transformation (Sec. 3.4.4) to obtain the GVI and SBI components from the four band Landsat MSS image for each of two dates. Pixels or groups of pixels (Malila preprocessed the imagery with the spatial clustering algorithm BLOB) are then classified according to the direction and magnitude of their signature change in the two-dimensional GVI-SBI feature space (Fig. 3-38). This choice of features and their vector interpretation is particularly useful for mapping changes in vegetative cover over time.

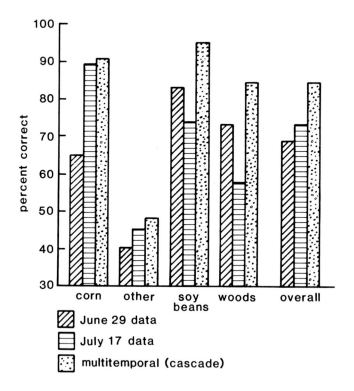

FIGURE 3-37. Comparison of classification accuracies from single-date and multitemporal cascade analysis (Swain, 1978b; ©1978 IEEE).

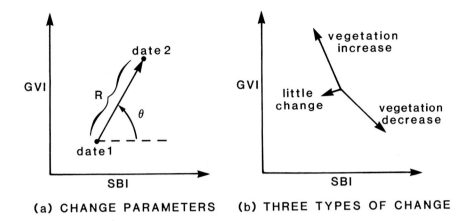

(a) CHANGE PARAMETERS   (b) THREE TYPES OF CHANGE

FIGURE 3-38. Change vector analysis.

Extension of the simple change detection technique described in Sec. 2.2.3 to multispectral images and multivariate
classification results in the multitemporal analysis procedure
of Fig. 3-39. The original double dimensionality of the two
multispectral images is reduced to the dimensionality of one
image by the differencing operation. Specification of training
sites for "change classes" may be either supervised or unsupervised depending on availability of information about the area
and the complexity of the changes between the two images. As in
the case of stacked multitemporal images, preprocessing must be
used to normalize the external factors of atmospheric scattering
and solar angle between the two images before they are differ-
enced. Spectral ratios may be calculated for each date and used
instead of the original images in the differencing step to automatically normalize for solar angle (Chavez et al, 1977).

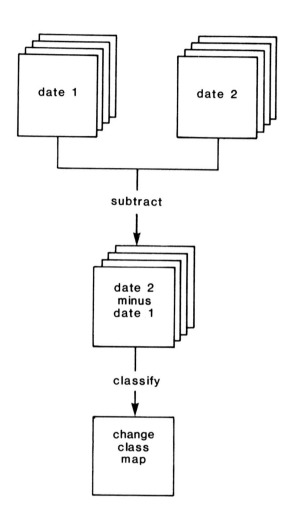

FIGURE 3-39. Change classification.

3.7.2 Digital Elevation Data

It is well-known that vegetation type and cover density, soils, and geology can be strongly correlated with surficial topography. The use of topographic data in classification has been limited, however, by the unavailability of the data in digital form. In the last few years, the Defense Mapping Agency and National Cartographic Information Center have produced digital topographic data for the entire United States from 1:250,000 scale maps. Furthermore, convenient and relatively inexpensive coordinate digitizers are now available, permitting the digitization of any conventional topographic contour map. Thus, these valuable data can now be incorporated into multispectral classifications (Hoffer and staff, 1975).

Figure 3-40a depicts a 3°-by-2° segment of digital elevation data for central Washington state; elevation is displayed simply as gray levels proportional to height above sea level. The bright area running through the center of the image is the Cascade Mountain range and darker areas extending into this region are valleys. Manual or automatic registration between this data and Landsat or other satellite images may be facilitated by first producing a *synthetic reflectance image* (Batson et al, 1975; Horn, 1981) as shown in Fig. 3-40b. This synthetic image is generated by specifying the desired solar elevation and azimuth angles and then calculating, at each pixel in the digital elevation data, the cosine of the angle between the local terrain surface normal vector and the solar vector. If a Lambertian reflectance model is assumed for the terrain, no further calculations are necessary to yield the final reflectance image. Two solar azimuth angles are shown in Fig. 3-40, both with a low solar elevation angle of ten degrees, to enhance topographic expression of faults in two directions (Schowengerdt and Glass, 1983). For registration purposes, the solar eleva-

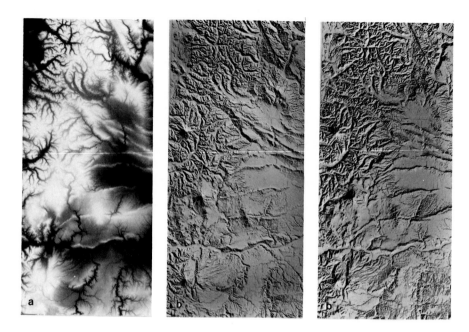

FIGURE 3-40.   Display of digital elevation data. (a) Elevation
image. (b) Synthetic reflectance image: left,
solar azimuth 45°; right, solar azimuth 135°.
(Schowengerdt and Glass, 1983; reprinted with permission from Geological Society of America Bulletin)

tion and azimuth of the synthetic image should be specified to
be equal to those for the satellite image. The synthetic image
may then be correlated numerically with the satellite image, or
control points may be located visually in both images. This
image-to-map registration technique is particularly valuable in
areas of high relief where few man-made objects are available
for use as control points.

Once the satellite image and elevation data are registered,
the latter may be incorporated into classification of the image
in several ways (Hutchinson, 1982). The most direct way is to

simply stack the elevation data with the multispectral image as an additional feature. Although simple, this *logical channel* approach (Strahler et al, 1978) contains subtle pitfalls. For example, a single training site for a surface cover class will probably contain very little variation in the elevation feature. Furthermore, certain classes such as water or bare soil may occur at virtually any elevation. Thus, a deliberate effort must be made to ensure that the full range of elevation occupied by each surface cover class is represented in the training data. This additional requirement on the number of training sites may be difficult to satisfy, however, because of limited verification data.

Strahler et al (1978) circumvented these problems by manually altering the means and variances of the elevation training samples to simulate the availability of more training samples in an interactive "tuning" process to achieve the highest possible classification accuracy. Fleming and Hoffer (1979) addressed the training sample problem by using a *topographic stratified random sample* approach in which elevation was stratified into relatively broad ranges and random pixel samples were selected within each stratum. The surface cover type of each sample was then determined and used to construct a topographic distribution model for each cover class. The class statistics developed from this model were then used for supervised training in a conventional single stage classification of the stacked multispectral and elevation data and in a two stage decision tree classification. There was little difference in accuracy between the two classifications, although the decision tree classification used considerably less computer time, as expected. Strahler et al (1978) also used a random sample of pixels to develop an elevation-dependent surface cover model. Rather than use the model to train the classifier, however, they used it to estimate *a priori* probabilities for spectral classes at different elevations.

Figure 3-41 depicts how topographic elevation might be used in a decision tree classification. It may be appropriate in certain situations to use elevation as the feature for the first stage followed by application of spectral features, or to mix spectral and elevation features in both stages. If the decision rules for the use of the elevation feature are not developed from statistical sampling but are defined deterministically by the analyst, they may be considered a type of *pre-classification stratification* or *post-classification sorting* (Hutchinson, 1982) depending on whether the elevation data is used before or after the spectral (or other) features. The latter procedure permits merging of classes, if desired, in addition to further discrimination.

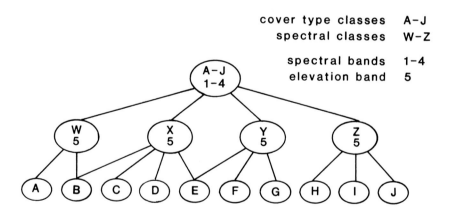

cover type classes   A-J
spectral classes   W-Z

spectral bands   1-4
elevation band   5

FIGURE 3-41. *Decision tree classifier with spectral and elevation features.*

Terrain slope and aspect features derived from digital elevation data may also be used in classification, although both Strahler et al (1978) and Fleming and Hoffer (1979) concluded that elevation was the best single discriminating feature in their forestry applications. The techniques described in this section may also be used to incorporate any other type of non-image spatially-distributed data into image classification (Missallati et al,1979). The specific approach used should not be selected arbitrarily, however, without evaluating which method is best suited to the application.

References

Addington, J.D., "A Hybrid Classifier Using the Parallelepiped and Bayesian Techniques", Proc. of the Am. Soc. of Photogrammetry Annual Meeting, March 9-14, 1975, pp. 772-784.

Anderberg, Michael R., Cluster Analysis for Applications, New York, Academic Press, 1973, 353 pp.

Badhwar, G.D., J.G. Carnes, and W.W. Austin, "Use of Landsat-Derived Temporal Profiles for Corn-Soybean Feature Extraction and Classification," Remote Sensing of Environment, Vol. 12, No. 1, March 1982, pp. 57-79.

Batson, R.M., Kathleen Edwards, and E.M. Eliason, "Computer-generated Shaded Relief Images," Journal of Research U.S. Geological Survey, Vol. 3, No. 4, 1975, pp. 401-408.

Bauer, Marvin E., Marilyn M. Hixson, Barbara J. Davis, and Jeanne B. Etheridge, "Crop Identification and Area Estimation by Computer-Aided Analysis of Landsat Data," Proc. of the Symposium on Machine Processing of Remotely Sensed Data, IEEE 77CH1218-7MPRSD, 1977.

Bevington, Philip R., Data Reduction and Error Analysis for the Physical Sciences, New York, McGraw-Hill, 1969, 336 pages.

Billingsley, Fred C., "Modeling Misregistration and Related Effects on Multispectral Classification," Photogrammetric Engineering and Remote Sensing, Vol. 48, No. 3, March 1982, pp. 421-430.

Blodget, H.W., F.J. Gunther, and M.H. Podwysocki, "Discrimination of Rock Classes and Alteration Products in South-western Saudi Arabia with Computer-Enhanced Landsat Data," NASA Technical Paper 1327, NASA Scientific and Technical Information Office, October, 1978.

Brayer, J.M., P.H. Swain and K.S. Fu, "Modeling of Earth Resources Satellite Data," Chapter 9 of Syntatic Pattern Recognition, Applications, K.S. Fu, ed., New York, Springer-Verlag, 1977.

Bryant, Jack, "On the Clustering of Multidimensional Pictorial Data," Pattern Recognition, Vol. 11, 1979, pp. 115-125.

Chavez, P. Jr., "Atmospheric, Solar, and MTF Corrections for ERTS Digital Imagery", Proc. Am. Soc. of Photogrammetry Fall Convention, October 1975.

Chavez, Pat S. Jr., G. Lennis Berlin, and William B. Mitchell, "Computer Enhancement Techniques of Landsat MSS Digital Images for Land Use/Land Cover Assessments," Proc. of the Sixth Annual Remote Sensing of Earth Resources Conf., U. of Tennessee Space Institute, March 1977.

Chavez, Pat S. Jr., Graydon L. Berlin, and Lynda B. Sowers, "Statistical Method for Selecting Landsat MSS Ratios," J. of Appl. Photographic Engineering, Vol. 8, No. 1, February, 1982, pp. 23-30.

Davis, John C., Statistics and Data Analysis in Geology, New York, John Wiley and Sons, 1973, 550 pages.

Davis, W.A., and F.G. Peet, "A Method of Smoothing Digital Thematic Maps", Remote Sensing of Environment, Vol. 6, No. 1, 1977, pp. 45-49.

Duda, R.D., and Hart, P.E. Pattern Classification and Scene Analysis, New York, John Wiley and Sons, 1973, 482 pp.

Duggin, M.J., "On the Natural Limitations of Target Differentiation by Means of Spectral Discrimination Techniques", Proc. Ninth International Symposium on Remote Sensing of Environment, Ann Arbor, Mich., April 1974, Vol. 1, pp. 499-516.

Eppler, W.G., "An Improved Version of the Table Look-Up Algorithm for Pattern Recognition", Proc. of the Ninth International Symposium on Remote Sensing of Environment, April 1974, Vol. 2, pp. 793-812.

Fleming, Michael D. and Roger M. Hoffer, "Machine Processing of Landsat MSS Data and DMA Topographic Data for Forest Cover Type Mapping," LARS Technical Report 062879, Purdue University, Laboratory for Applications of Remote Sensing, 1979.

Fukunaga, Keinosuke, Introduction to Statistical Pattern Recognition, New York, Academic Press, 1972, 369 pp.

Goetz, A.F.H., F.C. Billingsley, A.R. Gillespie, M.J. Abrams, R.L. Squires, E.M. Shoemaker, I. Luchitta and D.P. Elston, "Application of ERTS Images and Image Processing to Regional Geologic Problems and Geologic Mapping in Northern Arizona", Jet Propulsion Laboratory Technical Report 32-1597 prepared for NASA Contract 7-100, California Inst. of Technology, May 1975.

Goldberg, M., D. Goodenough and S. Shlien, "Classification Methods and Error Estimation for Multispectral Scanner Data", Proc. Third Canadian Symposium on Remote Sensing, September 1975, pp. 125-143.

Haralick, Robert M. and I. Dinstein, "A Spatial Clustering Procedure for Multi-Image Data," IEEE Transactions on Circuits and Systems, Vol. CAS-22, No. 5, May 1975, pp. 440-450.

Haralick, R.M., K. Shanmugam, and I. Dinstein, "Textural Features for Image Classification", IEEE Transactions on Systems, Man, and Cybernetics, Vol. SMC-3, No. 6, November 1973, pp. 610-621.

Hartigan, John A., Clustering Algorithms, New York, John Wiley Interscience, 1975, 351 pages.

Hay, Alan M., "Sampling Designs to Test Land-Use Map Accuracy", Photogrammetric Engineering and Remote Sensing, Vol. 45, No. 4, April, 1979, pp. 529-533.

Hoffer, R.M., and staff, "Computer-Aided Analysis of SKYLAB Multispectral Scanner Data in Mountainous Terrain for Land Use, Forestry, Water Resources, and Geologic Applications", LARS Information Note 121275, Purdue University, Laboratory for Applications of Remote Sensing, 1975.

Hord, R. Michael and William Brooner, "Land-Use Map Accuracy Criteria", Photogrammetric Engineering and Remote Sensing, Vol. 42, No. 5, May 1976, pp. 671-677.

Horn, B.K.P., "Hill Shading and the Reflectance Map," Proc. IEEE, Vol. 69, No. 1, 1981, pp. 14-47.

Horwitz, Harold M., Richard F. Nalepka, Peter D. Hyde and James P. Morgenstern, "Estimating the Proportions of Objects within a Single Resolution Element of a Multispectral Scanner," Proc. Seventh International Symposium on Remote Sensing of Environment, Ann Arbor, Mich., Vol. 2, May 1971, pp. 1307-1320.

Hulstrom, Roland L., "Spectral Measurements and Analyses of Atmospheric Effects on Remote Sensor Data", Scanners and Imagery Systems for Earth Observations, Proc. Soc. of Photo-optical Instrumentation Engineers, Vol. 51, August 1974, pp. 90-100.

Hutchinson, Charles F., "Techniques for Combining Landsat and Ancillary Data for Digital Classification Improvement," Photogrammetric Eng. and Remote Sensing, Vol. 48, No. 1, January 1982, pp. 123-130.

Jackson, Ray D., Robert J. Reginato, Paul J. Pinter, Jr., and Sherwood B. Idso, "Plant Canopy Information Extraction from Composite Scene Reflectance of Row Crops", Applied Optics, Vol. 18, No. 22, 15 November 1979, pp. 3775-3782.

Jackson, Ray D., "Spectral Indices in N-Space," Remote Sensing of Environment (in press 1983).

Jenson, Susan K. and Frederick A. Waltz, "Prinicpal Components Analysis and Canonical Analysis in Remote Sensing," Proc. Am. Soc. of Photogrammetry Annual Meeting, March 1979.

Kauth, R.J. and G.S. Thomas, "The Tasselled Cap -- A Graphic Description of the Spectral-Temporal Development of Agricultural Crops as Seen by Landsat," Proc. Symposium on Machine Processing of Remotely Sensed Data, IEEE 76CH 1103-1MPRSD, 1976, pp. 41-51.

Kauth, R.J., A. P. Pentland, and G.S. Thomas, "BLOB, an Unsupervised Clustering Approach to Spatial Preprocessing of MSS Imagery," Proc. of the Eleventh International Symposium on Remote Sensing of Environment, 1977, pp. 1309-1317.

Kettig, R.L. and D.A. Landgrebe, "Classification of Multispectral Image Data by Extraction and Classification of Homogeneous Objects," IEEE Trans. on Geoscience Electronics, Vol. GE-4, No. 1, January 1976, pp. 19-26.

Landgrebe, David A., "The Quantitative Approach: Concept and Rationale," pp. 1-20 in Remote Sensing: The Quantitative Approach, Swain and Davis, eds., New York, McGraw-Hill, 1978, 396 pages.

Lindenlaub, John C. and Shirley M. Davis, "Applying the Quantitative Approach," pp. 290-335 in Remote Sensing: The Quantitative Approach, Swain and Davis, eds., New York, McGraw-Hill, 1978, 396 pages.

Malila, William A., "Change Vector Analysis: An Approach for Detecting Forest Changes with Landsat," Proc. Symposium on Machine Processing Remotely Sensed Data, IEEE 80CH 1533-9MPRSD, 1980, pp. 326-335.

Malila, William A., Peter F. Lambeck, and Eric P. Crist, "Landsat Features for Agricultural Applications," Proc. of the Fourteenth International Symposium on Remote Sensing of Environment," 1980.

Markham, B.L. and J.R.G. Townshend, "Land Cover Classification Accuracy as a Function of Sensor Spatial Resolution," Proc. of the Fifteenth International Symposium on Remote Sensing of Environment, 1981, pp. 1075-1090.

Maxwell, Eugene L., "Multivariate Systems Analysis of Multispectral Imagery", Photogrammetric Engineering and Remote Sensing, Vol. 42, No. 9, September 1976, pp. 1173-1186.

Missallati, A., A.E. Prelat, and R.J.P. Lyon, "Simultaneous Use of Geological, Geophysical, and LANDSAT Digital Data in Uranium Exploration," Remote Sensing of Environment, Vol. 8, 1979, pp. 189-210.

Nalepka, Richard P. and Peter D. Hyde, "Classifying Unresolved Objects from Simulated Space Data", Proc. Eighth International Symposium on Remote Sensing of Environment, Environmental Research Institute of Michigan, Ann Arbor, Mich., Vol. 2, October 1972, pp. 935-949.

Pitts, D.E., W.E. McAllum and A.E. Dillinger, "The Effect of Atmospheric Water Vapor on Automatic Classification of ERTS Data," Proc. Ninth International Symposium on Remote Sensing of Environment, Ann Arbor, Mich., April 1974, pp. 483-498.

Potter, J.F. and M.A. Mendlowitz, "On the Determination of Haze Levels from Landsat Data," Proc. Tenth International Symposium on Remote Sensing of Environment, Vol. 2, October 1975, pp. 695-703.

Pratt, William K., Digital Image Processing, New York, John Wiley and Sons, 1978, 750 pp.

Richardson, A.J. and C.L. Wiegand, "Distinguishing Vegetation from Soil Background Information," Photogrammetric Engineering and Remote Sensing, Vol. 43, No. 12, December 1977, pp. 1541-1552.

Rouse, J. W. Jr., R.H. Hass, J.A. Schell, and D.W. Deering, "Monitoring Vegetation Systems in the Great Plains with ERTS," Proc. Third ERTS Symposium, NASA SP-351, December 1973, Vol. I, pp. 309-317.

Salvato, Pete Jr., "Iterative Techniques to Estimate Signature Vectors for Mixture Processing of Multispectral Data," Proc. Symposium on Machine Processing of Remotely Sensed Data, IEEE 73CHO834-2GE, 1973, pp. 3B:48-62.

Schowengerdt, Robert A. and Charles E. Glass, "Digitally Processed Topographic Data for Regional Tectonic Evaluations," Geological Society of America Bulletin, Vol. 94, April 1983, pp. 549-556.

Slater, Philip N., Remote Sensing - Optics and Optical Systems, Reading, Mass., Addison-Wesley, 1980, 575 pp.

Strahler, Alan H., Thomas L. Logan and Nevin A. Bryant, "Improving Forest Cover Classification Accuracy from Landsat by Incorporating Topographic Information," Proc. Twelfth International Symposium on Remote Sensing of Environment, Ann Arbor, Mich., 1978, pp. 927-942.

Swain, Philip H., "Fundamentals of Pattern Recognition in Remote Sensing," pp. 136-187 in Remote Sensing: The Quantitative Approach, Swain and Davis, eds., New York, McGraw-Hill, 1978, 396 pages.

Swain, Philip H., "Bayesian Classification in a Time-varying Environment," LARS Technical Report 030178, Purdue University, Laboratory for Applications of Remote Sensing, 1978b.

Swain, Philip H. and Hans Hauska, "The Decision Tree Classifier: Design and Potential," IEEE Trans. on Geoscience Elec tronics, Vol. GE-15, No. 3, July 1977, pp. 142-147.

Swain, Philip H. and Shirley M. Davis, eds., Remote Sensing: The Quantitative Approach, New York, McGraw-Hill, 1978, 396 pp.

Thompson, David R. and Oscar A. Wehmanen, "Using Landsat Digital Data to Detect Moisture Stress," Photogrammetric Eng. and Remote Sensing, Vol. 45, No. 2, February 1979, pp. 201-207.

Todd, William J., Dale G. Gehring, and Jon F. Haman, "Landsat Wildland Mapping Accuracy," Photogrammetric Engineering and Remote Sensing, Vol. 46, No. 4, April 1980, pp. 509-520.

Tucker, C.J., "Use of Near Infrared/Red Radiance Ratios for Estimating Vegetation Biomass and Physiological Status", Proc. Eleventh International Symposium on Remote Sensing of Environment, 1977.

Tucker, Compton J., "Red and Photographic Infrared Linear Combinations for Monitoring Vegetation," Remote Sensing of Environment, Vol. 8, 1979, pp. 127-150.

Tucker, Compton J. and Lee D. Miller, "Soil Spectra Contributions to Grass Canopy Spectral Reflectance," Photogrammetric Engineering and Remote Sensing, Vol. 43, No. 6, June 1977, pp. 721-726.

van Genderen, J.L., B.F. Lock and P.A. Vass, "Remote Sensing: Statistical Testing of Thematic Map Accuracy," Remote Sensing of Environment, Vol. 7, 1978, pp.3-14.

Wecksung, G.W. and J.R. Breedlove Jr., "Some Techniques for Digital Processing, Display, and Interpretation of Ratio Images in Multispectral Remote Sensing", Proc. Soc. of Photo-optical Instrumentation Engineering, Vol. 119, Applications of Digital Image Processing, 1977, pp. 47-54.

Wharton, Stephen W., "A Context-Based Land-Use Classification Algorithm for High-Resolution Remotely Sensed Data," J. of Appl. Photographic Engineering, Vol. 8, No. 1, February 1982, pp. 46-50.

Wiersma, D.J. and D. Landgrebe, "The Use of Spatial Characteristics for the Improvement of Multispectral Classification of Remotely Sensed Data", Proc. Symposium on Machine Processing of Remotely Sensed Data, IEEE 76CH1103-1 MPRSD, 1976.

Remote Sensing and Image Processing Bibliography

This appendix is a reasonably current and comprehensive guide to the scientific literature in these two fields. No claims are made about the completeness of this collection; many items that would be part of a complete bibliography have been purposely excluded because of minimal relevance to the topics in this book. On the other hand, some important topics that are not covered in the text, such as image coding, are included here. All the following lists are in chronological order to give a perspective on history and to permit the reader to begin further reading with the most recent literature.

A.1 Books

Image Processing and Classification

Andrews, H.C., Computer Techniques in Image Processing, New York, Academic Press, 1968.

Andrews, H.C., Introduction to Mathematical Techniques in Pattern Recognition, New York, Wiley-Interscience, 1972, 242 pp.

Fukunaga, Keinsoke, Introduction to Statistical Pattern Recognition, New York, Academic Press, 1972, 369 pp.

Duda, R.D. and P.E. Hart, Pattern Classification and Scene Analysis, New York, John Wiley and Sons, 1973, 482 pp.

Tou, J.T. and R.C. Gonzalez, Pattern Recognition Principles, 2nd ed., Reading, Mass., Addison-Wesley, 1975.

Rosenfeld, Azriel and Avinash C. Kak, Digital Picture Processing, New York, Academic Press, 1976, 457 pp.

Aggarwal, J.K., Richard O. Duda, and Azriel Rosenfeld, eds., Computer Methods in Image Analysis, IEEE Press, 1977, 466 pp. (reprints).

Agrawala, Ashok K., ed., Machine Recognition of Patterns, IEEE Press, 1977, 463 pp. (reprints).

Andrews, Harry C. and B.R. Hunt, Digital Image Restoration, Englewood Cliffs, New Jersey, Prentice-Hall, 1977.

Gonzalez, Rafael C. and Paul Wintz, Digital Image Processing, Reading, Mass., Addison-Wesley, 1977, 431 pp.

Andrews, Harry C., ed., Digital Image Processing, IEEE Computer Society, 1978, 728 pp. (reprints).

Bernstein, Ralph, ed., Digital Image Processing for Remote Sensing, IEEE Press, 1978, 484 pp. (reprints).

Pratt, William K., Digital Image Processing, New York, John Wiley and Sons, 1978, 750 pp.

Castleman, Kenneth R., Digital Image Processing, Englewood Cliffs, New Jersey, Prentice-Hall, Inc., 1979, 429 pp.

Hall, Ernest L., Computer Image Processing and Recognition, New York, Academic Press, 1979, 584 pp.

Moik, Johannes G., Digital Processing of Remotely Sensed Images, NASA SP-431, Washington, D.C., U.S. Government Printing Office, 1980, 330 pp. (out of print, 1983).

Ballard, Dana H., and Christopher M. Brown, Computer Vision, Englewood Cliffs, New Jersey, Prentice-Hall, 1982, 523 pp.

Cracknell, A.P., ed., Computer Programs for Image Processing of Remote Sensing Data, Dundee, Scotland, University of Dundee, 1982, 91 pp.

Foley, James D. and Andries van Dam, Fundamentals of Interactive Computer Graphics, Reading, Mass., Addison-Wesley, 1982, 690 pp.

Green, William, Digital Image Processing - A Systems Approach, New York, Van Nostrand Reinhold, 1982, 288 pp.

Hord, R. Michael, Digital Image Processing of Remotely Sensed Data, New York, Academic Press, 1982, 256 pp.

Monmonier, Mark S., Computer-Assisted Cartography Principles and Prospects, Englewood Cliffs, New Jersey, Prentice-Hall, 1982, 214 pp.

Pavlidis, Theo, Algorithms for Graphics and Image Processing, Rockville, Maryland, Computer Science Press, 1982, 416 pp.

Remote Sensing

Reeves, Robert G., editor-in-chief, Manual of Remote Sensing, First Edition, Falls Church, Virginia, American Society of Photogrammetry, 1975, 2100 pp.

Sabins, Floyd F. Jr., Remote Sensing - Principles and Interpretation, San Francisco, W.H. Freeman and Co., 1978, 426 pp.

Swain, P.H. and S.M. Davis, eds., Remote Sensing - The Quantitative Approach, New York, McGraw-Hill, 1978, 396 pp.

Lillesand, Thomas M. and Ralph W. Kiefer, Remote Sensing and Image Interpretation, New York, John Wiley and Sons, 1979, 612 pp.

Siegal, Barry S. and Alan R. Gillespie, eds., Remote Sensing in Geology, New York, John Wiley and Sons, 1980, 702 pp.

Slater, Philip N., Remote Sensing - Optics and Optical Systems, Reading, Mass., Addison-Wesley, 1980, 575 pp.

Short, Nicholas M., The Landsat Tutorial Workbook, NASA RP-1078, Washington, D.C., U.S. Government Printing Office, 1982, 553 pp.

Colwell, Robert N., editor-in-chief, Manual of Remote Sensing, Second Edition, Falls Church, Virginia, American Society of Photogrammetry, 1983, 2400 pp.

A.2 Topical Papers

Stockham, T.G. Jr., "Image Processing in the Context of a Visual Model," Proc. IEEE, Vol. 60, 1972, pp. 828-842.

Hunt, B.R., "Digital Image Processing," Proc. IEEE, Vol. 63, 1975, pp. 693-708.

Hunt, B.R. and J.R. Breedlove, "Sample and Display Considerations in Processing Images by Digital Computer," IEEE Trans. Computers, Vol. 24, 1975, pp. 848-853.

Anderson, James R., Ernest E. Hardy, John T. Roach, and Richard E. Witmer, "A Land Use and Land Cover Classification System for Use with Remote Sensor Data," Washington, D.C., U.S. Government Printing Office, U.S. Geological Survey Professional Paper 964, 1976, 28 pp.

Reader, Clifford and Larry Hubble, "Trends in Image Display Systems," Proc. IEEE, Vol. 69, No. 5, May 1981, pp. 606-614.

Stoffel, J.C. and J.F. Moreland, "A Survey of Electronic Techniques for Pictorial Image Reproduction," IEEE Trans. on Communications, Vol. COM-29, No. 12, December 1981, pp. 1898-1925.

Nagy, George, "Optical Scanning Digitizers," Computer, IEEE Computer Society, Vol. 16, No. 5, May 1983, pp. 13-24.

A.3 Literature Surveys

Rosenfeld, Azriel, "Picture Processing," Computer Graphics and Image Processing, 1972 (Vol. 1, 1972, pp. 394-416), 1973 (Vol. 3, 1974, pp. 178-194), 1974 (Vol. 4, 1975, pp. 133-155), 1975 (Vol. 5, 1976, pp. 215-237), 1976 (Vol. 6, 1977, pp. 157-183), 1977 (Vol. 7, 1978, pp. 211-242), 1978 (Vol. 9, 1979, pp. 354-393), 1979 (Vol. 13, 1980, pp. 46-79), 1980 (Vol. 16, 1981, pp. 52-89), 1981 (Vol. 19, 1982, pp. 35-75).

Kanal, Laveen, "Interactive Pattern Analysis and Classification Systems: A Survey and Commentary," Proc. IEEE, Vol. 60, Oct. 1972, pp. 1200-1215.

Nagy, George, "Digital Image Processing Activities in Remote Sensing for Earth Resources," Proc. IEEE, Vol. 60, Oct. 1972, pp. 1177-1200.

Andrews, Harry, C., "Digital Image Restoration: A Survey," Computer, May 1974, pp. 36-45.

Davis, Larry S., "A Survey of Edge Detection Techniques," Computer Graphics and Image Processing, Vol. 4, 1975, pp. 248-270.

Kanal, Laveen, "Patterns in Pattern Recognition: 1968-1974," IEEE Trans. Inform. Theory, Vol. IT-20, Nov. 1974, pp. 697-722.

Pratt, William K., "Survey and Analysis of Image Coding Techniques," Proc. of the Society of Photo-optical Instrumentation Engineers, Vol. 74, Image Processing, 1976, pp. 178-184.

Netravali, A.N. and J.O. Limb, "Picture Coding: A Review," Proc. IEEE, Vol. 68, No. 3, 1980, pp. 366-406.

Jain, Anil K., "Image Data Compression: A Review," Proc. IEEE, Vol. 69, No. 3, March 1981, pp. 349-389.

A.4 Special Journal Issues

Proc. IEEE, Special Issue on Digital Image Processing, Vol. 60, No. 7, July 1972.

Proc. IEEE, Special Issue on Digital Pattern Recognition, Vol. 60, No. 10, October 1972.

Computer, IEEE Computer Society, Special Issue on Digital Image Processing, Vol. 7, No. 5, May 1974.

Proc. IEEE, Special Issue on Pattern Recognition and Image Processing, Vol. 67, No. 5, May 1979, pp. 707-859.

Proc. IEEE, Special Issue on Image Processing, Vol. 69, No. 5, May 1981, pp. 497-672.

Journal of Applied Photographic Engineering, Special Issues on Remote Sensing and Digital Image Processing, Vol. 8, Nos. 1 and 3, February and June, 1982.

Computer, IEEE Computer Society, Special Issue on Computer Architectures for Image Processing, Vol. 16, No. 1, January 1983.

APPENDIX B

Digital Image Data Formats

One of the first problems encountered by anyone who becomes involved in writing image processing software is manipulation and conversion of a wide variety of data formats. This generally nonproductive activity often consumes a disproportionate share of personnel and computer resources. In this appendix, we describe the basic characteristics of digital image formats and storage in an attempt to remove some of the mystery from this aspect of image processing.

B.1 Bits and Pixels

The gray level of each pixel of a digital image is recorded and stored as a finite number of *bits*. A bit is the elementary unit of binary computing; it has only two possible values, either 1 (the bit is "on") or 0 (the bit is "off"). By using a string of bits we can represent an arbitrarily large number within the limit of the computer *word length*, which is the maximum number of bits available to represent a number in a given computer. Typical word lengths are 16, 32, and 60 bits. Longer word lengths permit greater precision in number representation and calculations and reduce concern about data overflow in calculations.

A simple example of the binary representation of numbers is shown in Fig. B-1. If there are k bits/pixel, a total of 2^k

bit map $b_3\ b_2\ b_1$	gray level
0 0 0	0
0 0 1	1
0 1 0	2
0 1 1	3
1 0 0	4
1 0 1	5
1 1 0	6
1 1 1	7

$$\text{Gray level} = \sum_{i=1}^{k} b_i 2^{i-1}$$

FIGURE B-1. Binary representation of gray levels (k = 3 bits/pixel).

gray levels are possible, over the range 0 to $2^k - 1$ (Table B-1). If k equals 8, the group of bits representing each pixel is called a *byte*, a common unit for processing image data. Note that bits are conventionally numbered from right to left, i.e., from the *least significant bit* (LSB) to the *most significant bit* (MSB).

Table B-1. Typical Radiometric Quantization Parameters
for Digital Imagery

k (bits/pixel)	number of gray levels	minimum gray level	maximum gray level
5	32	0	31
6	64	0	63
7	128	0	127
8	256	0	255
9	512	0	511

Most minicomputers and many large machines have system software that permits direct access to subunits of a word and, consequently, single bytes of data can be manipulated directly. Some large computers, such as the CDC Cyber 170 series, however, can only process full words consisting of 60 bits. Image data consisting of 1 byte/pixel therefore must be *packed* within each computer word for maximum storage efficiency (Fig. B-2). The data then must be *unpacked* to a 1 pixel/word format for processing, and repacked to efficiently store the results. These data reformatting steps, of course, require additional computer time over and above that required for processing.

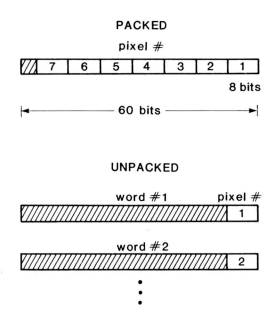

FIGURE B-2. *Packed and unpacked pixel data for CDC Cyber 170 series computers.*

B.2 Records and Files

The spatially-ordered nature of images leads naturally to a data storage format that preserves at least some of that intrinsic order. Thus, each line of image pixels is usually stored as a *physical record*. A record is a stream of data that is read or written with a single program command. The total set of records constituting an image is called a *file*.

Multispectral images require special consideration because they contain separate images that may be treated logically as a single image. They are stored in one of three *interleaved* formats: band-interleaved by pixel (BIP), band-interleaved by line (BIL), or band sequential (BSQ). These formats are diagrammed in Fig. B-3.[1] Each of these formats is particularly suited to certain types of processing. If the processing is a pixel-by-pixel multispectral classification, for example, the BIP format is convenient because the pixel gray levels in each band are stored contiguously within a data record. If the processing is only on single bands from the multispectral image, however, the BSQ format is most attractive because it minimizes the amount of data that must be read to access a single band. The BIL format represents a good compromise of efficiency and convenience for general application and is probably used more widely than either of the other formats.

Image files that are produced for general distribution, such as Landsat scenes, usually contain header and/or trailer records (at the beginning and end of image files, respectively) for ancillary data such as image identification, the number of pixels/line (or bytes/record), the number of lines, the number of bands, radiometric calibration data, etc. These data usually are stored in a format that is different from that used for the image data, and is described in documentation accompanying the data tapes.

B.3 Tapes and Disks

Digital data is stored on a magnetic medium, such as mylar tape coated with a thin film of iron oxide or thin metallic disks. Magnetic tape and disk media can be written on, read from, and rewritten many times. Optical disk storage is only beginning to be used and is not available in most computer facilities at this time.

---

[1]A variation on the BIP format, known as BIP2, was used for several years for Landsat MSS data (Holkenbrink, 1978). The BIP2 format consisted of *pairs* of pixels from each band interleaved within each record.

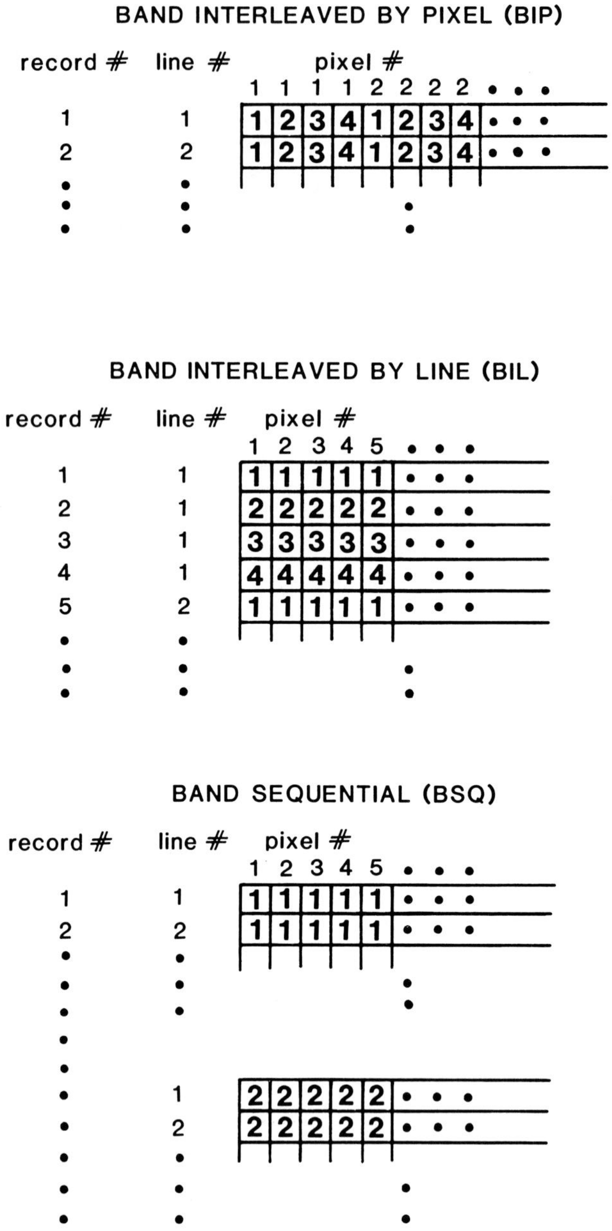
BAND INTERLEAVED BY PIXEL (BIP)
record #
line #
pixel #
1 1 1 1 2 2 2 2
1 2 3 4 1 2 3 4
BAND INTERLEAVED BY LINE (BIL)
record #
line #
pixel #
1 2 3 4 5
BAND SEQUENTIAL (BSQ)
record #
line #
pixel #
1 2 3 4 5

FIGURE B-3. Three common formats for digital multispectral images. The boldface numbers in the squares denote the band number for a 4-band image.

Magnetic tape is a *sequential* storage medium because access
to a particular location on the tape requires physically passing
over all the data recorded before that point. Bits are recorded
on a tape in nine tracks that parallel the length of the tape
(seven-track tapes have become virtually obsolete). The
arrangement of these tracks on the tape is depicted in Fig.
B-4. Nine tracks are convenient for image data because each
pixel can be stored as a byte *across* the tape, reserving the
ninth bit for parity checking (this "parity bit" is turned on or
off automatically by the computer hardware to maintain a
consistent odd or even number of bits across the tape and serves
as a check for magnetically bad areas on the tape). The density
of bits *along* the tape is commonly 800, 1600, or 6250 bits/inch
(315, 630, or 2460 bits/cm) and determines the amount of data
that can be stored on the tape. For nine-track tapes and
1 byte/pixel, this bit density is also equal to the pixel
density. Note that the number of tracks and bit density are *not*

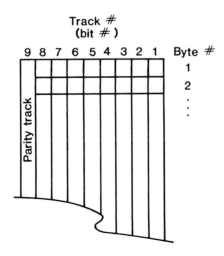

FIGURE B-4. Data format on nine-track tapes.

a physical characteristic of the tape, but are determined by the tape drive that is used to write data onto the tape.

Physical records on a tape are separated by end-of-record (EOR) marks.   Files are separated by end-of-file (EOF) marks. These "marks" consist of a blank gap (EOR) or a special bit pattern (EOF) of specified (universal) length and permit skipping of records or files without actually reading the intervening data into memory.   For the BIP and BIL formats, each multispectral image is separated from others on the same tape by an EOF; for the BSQ format each *band* is separated from others in the same multispectral image by an EOF; multiple scenes also are separated by EOFs.

The logical structure described above for tapes is carried over to storage on magnetic disks.   However, disks are *random* access media and therefore, although a complete image line is stored as a complete record in one location on the disk, the records that constitute a full image are not necessarily contiguous.   Data access from disk is normally faster than from tape because the data transfer rate is higher and individual records of data can be read without reading sequentially through "preceding" data as required with a tape.   Typical storage capacities for tapes and disks are summarized in Table B-2.

Table B-2.   Approximate Storage Capacities
for Magnetic Tapes and Disks

Magnetic tape (length 2400 ft (730 m)):

Density [bits/inch (bits/cm)]	Storage Capacity (M bytes)
800 (315)	23
1600 (630)	46
6250 (2460)	180

Table B-2--*Continued*

Magnetic disk:

Type	Storage Capacity (M bytes)
flexible	1
hard small	10
hard medium	80
hard large	300

The Table Look-Up Algorithm and Interactive Image Processing

There are several references in Chapters 2 and 3 to the use of look-up tables for efficient implementation of image processing or classification algorithms. Although the concept of a look-up table (LUT) is simple, its widespread use in software and recent applications in interactive hardware warrant the brief description in this appendix.

C.1 The Table Look-Up Algorithm

Contrast enhancement (Sec. 2.2) involves the transformation of each pixel's gray level into a new gray level. For the simplest type of enhancement, the transformation function is the same for every pixel in a particular image, and therefore may be written as

$$GL' = T(GL) \qquad\qquad (C-1)$$

where GL and GL' are the input and output gray levels, respectively, and T is the transformation function. Since both GL and GL' are discrete, quantized variables, T must also be discrete as shown in Fig. C-1.

In virtually all high level programming languages, such as Fortran, a discrete function like T(GL) may be defined as an

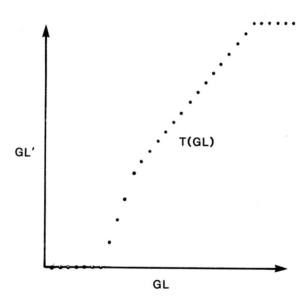

FIGURE C-1. An example discrete gray level transformation.

array with an integer index GL. If the same function T is to be used for all pixels in an image, it may be calculated for every possible value of GL *before* processing the image and stored during program execution as an array. An image with 8 bits per pixel would require a table of 256 values. To process the image, each pixel's gray level is simply used as the index in the table T. The corresponding value of T is then the output gray level. No calculations are necessary to perform the operation, only the indexing of a location in the table is required.

LUTs can be used for other purposes, such as storage of classification decision boundaries as described in Sec. 3.5.1. For the classification of K-dimensional data, the LUTs must also be K-dimensional. Again, a significant savings in computation time may be realized with the use of LUTs, particularly for complex classifiers such as the maximum-likelihood algorithm. The tables must be reset, however, if there is *any* change in the training data, a common circumstance in the process of refining a supervised classification. Thus, LUTs are ideally implemented in hardware form as described in the next section.

C.2 Interactive Processing

The use of a hardware LUT in an interactive monochrome image display is shown in Fig. C-2. The digital image is stored in the semiconductor refresh memory and completely read out every 1/30 second and displayed. Between the image memory and the CRT display is a semiconductor device that performs a discrete transformation on the digital data before display. Operator input is supplied to this LUT to control the form of the gray level transformation.

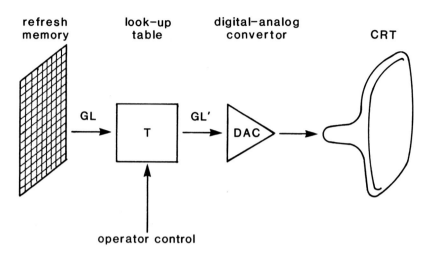

FIGURE C-2. Simplified schematic of an interactive monochrome CRT display.

The control of the LUT is made interactive by making the gray level transformation T a simple parametric form such as

$$GL' = T(GL) = aGL + b \qquad (C-2)$$

In this case, T is a linear contrast stretch with a gain factor a and a bias factor b. The parameters a and b are now linked directly to an operator-controlled x-y pointer, such as a CRT cursor or a graphics tablet and electronic pen. For example, the gain a may be made proportional to the y position of the pointer and the bias b may be made proportional to the x position of the pointer. Examples of how this linkage operates are shown in Fig. C-3.

Because the image is redisplayed, via the LUT, every 1/30 second, the operator can move the cursor or other pointer rapidly and view the resulting contrast-enhanced image instantaneously. This immediate feedback promotes not only efficient processing but also a subjective sense of operator

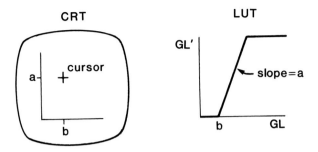

(a) PARAMETERS

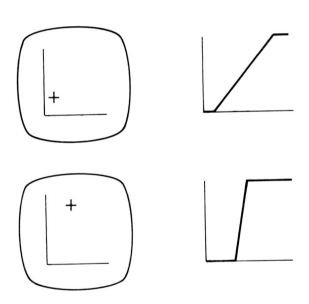

(b) EXAMPLES

FIGURE C-3. *Interactive implementation of a two-parameter contrast stretch.*

control that is impossible to achieve with a batch processing
system.   The   operator   can   experiment   with   the   contrast
transformation   until   the   desired   enhancement   is   achieved,
usually in less than a minute.

APPENDIX D

Examination Questions

This appendix contains a sampler of examination questions that I have asked over several years of teaching the material in this book. Although some questions require only a recall of appropriate sections in the text, many are designed to deduce the student's *understanding* of the concepts presented in the course. This collection serves as a starting point from which an instructor can develop questions that reflect individual emphasis and style.

D.1 Gray Levels, Histograms and Contrast Manipulation

(1) a. If a digital image has pixels quantized to 6 bits, how many gray levels are possible? What are they?

 b. Show which bits are "on" and which are "off" for a gray level of 23.

 c. If a Landsat MSS image typically has a range of 30 gray levels, why are 64 gray levels available?

 d. Why would we desire to have more bits representing pixel values in a computer than are used for the original pixel values? Illustrate with an example.

(2) a. What is a histogram of image gray levels? How is it usually normalized?

 b. Why is a histogram useful for changing image contrast? If we didn't have the image histogram, what parameters of the image could be used to adjust its contrast?

(3) a. It is sometimes said that contrast stretching increases the "information content" of an image. Do you agree or disagree with that statement and why?

 b. We have an image with the following histogram:

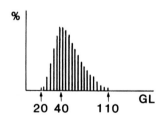

Sketch the GL transformation curves required to:

   i. stretch the image min and max GL to 0 and 255, respectively, with no GL saturation.

   ii. stretch the image in the same way as in part (i) but with some GL saturation at each end in the output.

iii. stretch the image as in part (i) but also place
      the output image's histogram peak at 128.

c. How could you equalize the image histogram and produce
    a negative image with *one* transformation?

D.2 Spatial Filtering, Fourier Transforms and Noise Suppression

(1) a. Suppose an image has the following histogram:

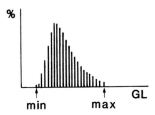

The image is filtered separately with each of the
following high-pass PSF filters:

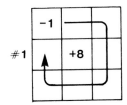   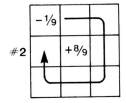

Sketch the two filtered image histograms on the same
graph.

b. Repeat part (a) for the following two low-pass PSF
    filters:

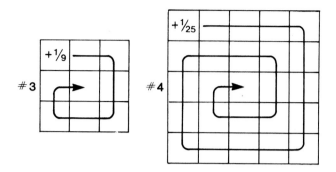

(2) Suppose we filter an image *twice*, each time with a 3x3 low-pass PSF filter:

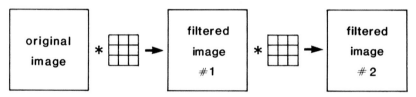

If we want to accomplish the same result with *one* filtering step, how big should the PSF window be?

(3) A processed image is given by,

$$i' = i * PSF$$

where, in this case, the volume under the PSF is one.

a. Show that a linear transformation (y = ax + b) of i, followed by convolution, yields the same result as when the same transformation is applied to i' directly.

b. Why, then, is the transformation normally done *after* the spatial filtering, for the purpose of display?

(4) a. Why are Fourier transforms useful in an analytic sense?

b. Why are Fourier transforms useful in a digital application?

c. Sketch both a high and low pass filter in the Fourier domain. What is a simple way to implement each in the spatial domain? Derive the number of operations (one

operation being an add, a multiply, etc.) required per
pixel to perform this simple low-pass operation with a
straight-forward calculation and with the more
efficient box-filter algorithm.

(5) a. Why is the Fourier transform useful for detecting
         periodic noise in an image?

     b. Given an image with periodic, stationary noise, how are
         Fourier transforms used to remove the noise?

D.3 Geometric Processing

(1) a. What is a "control point" (CP) for geometric
         processing?

     b. What are the characteristics of a "good" CP?

     c. If a CP is good for registration of two multitemporal
         images, is it also necessarily good for registration of
         the images to a map? Why or why not?

(2) a. Suppose we know there is only a shift and scale change
         between two images. The transformation polynomial is
         therefore

$$x = a_0 + a_1 x'$$

$$y = b_0 + b_1 y'$$

We have measured two CPs in both images to have the
following coordinates:

CP#1: $(x_1, y_1) = (10, 20)$ $(x_1', y_1') = (15, 18)$

CP#2: $(x_2, y_2) = (16, 28)$ $(x_2', y_2') = (17, 20)$

What is: i. the relative scale?
          ii. the relative shift?
between the two images in both the x and y directions.

     b. What happens if we have measured more than two CPs in
         part (a) and wish to use them all because some may be
         in error?

(3) We are interpolating an image in order to change its
     geometry. The position of an output pixel to be calculated
     is shown at X in the following figure:

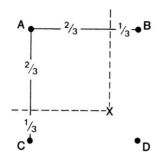

A, B, C, and D are the values of the four surrounding input pixels. What is the output pixel value for:

a. nearest neighbor interpolation?

b. bilinear interpolation?

(4) Suppose we have a Landsat image and a map of the same area. We want to process the image to make it overlay the map as accurately as possible. Describe all the steps of the necessary processing.

D.4 Multifeature Analysis

(1) What is a "pixel vector" for a three band multispectral image? What are the components of the vector along each axis?

(2) a. We have a three band multispectral image with the scattergrams between band 1 and band 2 and between band 1 and band 3 as follows:

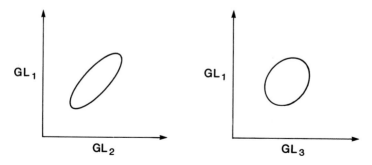

We now subtract band 2 from band 1 and band 3 from band 1, pixel-by-pixel. Sketch the gray level histograms of the two difference images below:

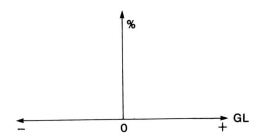

b. Explain the difference between the two histograms in terms of the correlation between band 1 and the other two bands.

(3) a. What does the term "feature" mean in classification?

b. Give three examples of different types of features.

c. What are the elements along the diagonal of the covariance matrix? What does it mean if all the off-diagonal elements are zero?

d. What is the difference between a principal and a canonical feature transformation?

D.5 Classification

(1) a. Describe the K-means clustering algorithm.

b. What criterion is used to stop the iteration of the algorithm?

(2) Two class feature distributions are shown below:

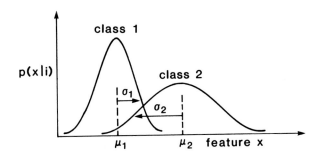

a. Define a simple quantity which expresses the "separability" of the two classes in terms of their means, μ_1 and μ_2 and standard deviations, σ_1 and σ_2. The measure of separability should always be positive and unitless.

b. Can you think of cases where the measure of separability defined in part (a) is misleading?

(3) a. Show the maximum-likelihood and nearest-mean (mininum-distance) decision boundaries on the following graph.

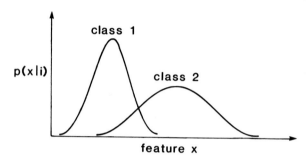

b. Which classifier results in smaller total error?

c. Which classifier results in more accurate classification of pixels belonging to class 1? Does it under- or over-estimate the total area (number of pixels) classified into class 1. Why?

(4) What is the difference between supervised and unsupervised training?

(5) Suppose you are given a magnetic tape containing a digital, multispectral image and are told to produce a vegetation map from the image, with a summary table of the area in km^2 that is vegetated within the scene. You are initially given no other information about the image data.

a. What questions about the image data will you have to ask and why?

b. You produce the requested classification map for the (paying) customer in part (a). Wanting to get his money's worth, he now wants to know how accurate the map is! Briefly describe three ways to estimate the accuracy and note the relative advantages/disadvantages of each approach.

D.6 Processing Considerations

(1)   Suppose we have an N x N image and want to rotate it 45°
       (either clockwise or counterclockwise). If the computer
       program retains in central memory at any given time the
       required number of input image lines to produce a given
       output line, what is the maximum amount of memory (number
       of pixels) required for rotating this image? Remember to
       include enough memory for one output image line.

(2)   Since most images are too big to store in the central
       memory of a computer, they are commonly stored on a
       magnetic disk. Many image processing programs read one or
       more image lines into memory from the disk and write the
       processed lines out onto the disk, one line at a time.
       Several lines of image data can usually be held in memory
       at once, depending of course on the length of the lines and
       amount of memory available.

       a.   If we have an N x N pixel image on disk, how long does
             it take to do a linear contrast stretch of the form
             $GL' = aGL + b$?

       b.   How long does it take to convolve the same image in the
             spatial domain with a W x W filter (with arbitrary
             weights)? Ignore complications at the boundaries of
             the image.

       c.   If your computing charges are based on the product of
             the total time and the amount of memory used during
             that time, what are your charges in each case above?

Index